Instrumental Insemination
of
Honey Bee Queens

Pictorial Instructional Manual

HARRY H. LAIDLAW, JR.

Professor of Entomology Emeritus

Apiculturist Emeritus

University of California, Davis

DADANT & SONS • HAMILTON, ILLINOIS

Northern Bee Books

Instrumental Insemination
of
Honey Bee Queens

Pictorial Instructional Manual

HARRY H. LAIDLAW, JR.

Professor of Entomology Emeritus

Apiculturist Emeritus

University of California, Davis

DADANT & SONS • HAMILTON, ILLINOIS

Northern Bee Books

ISBN 978-1-908904-32-4

Published by Northern Bee Books, 2013
Scout Bottom Farm
Mytholmroyd
Hebden Bridge
HX7 5JS (UK)

D&P Design and Print
Worcestershire

Printed by Lightning Source, UK

CONTENTS

PREFACE

Instrumental, or artificial, insemination of honey bee queens has reached near perfection. Injury to queens is a rarity, and successful insemination is virtually 100%. When drones are properly reared and matured so that most of them yield semen, an average of ten queens inseminated per hour per operator is not unusual.

This pleasant status owes its existence to the research and development efforts of several men, beginning with Watson and continuing to the present, who have made their contributions known by publication in scientific papers.

It seems appropriate, now, to prepare a teaching manual that in large measure will serve in lieu of a physically present instructor for those who wish to learn to artificially inseminate queen honey bees. This pictorial instructional manual leads the student step-by-step through the entire process of artificial insemination whether the instrument used is a Mackensen or a Laidlaw type.

Genetics and breeding of the honey bee are purposely omitted, as are all beekeeping operations, even those associated with queen rearing and queen care. Those subjects are extraneous in this manual and should be covered elsewhere.

I wish to express my appreciation to Jerry Marston, Research Associate, for supplying some of the queens used for photographing.

Special thanks are due Drs. Otto Mackensen, and F. Ruttner for their assistance in ensuring that the Synopsis is accurate and includes the contributions pertinent to the technique and instruments of artificial insemination as they are today, and to Howard Veatch for supervising the publication of this manual.

I am particularly grateful to my wife, Ruth, for her help in making the photographs and for doing all of the necessary typing.

Harry H. Laidlaw, Jr.

Davis, California
November 12, 1976

HARRY H. LAIDLAW, JR., Ph.D.

SYNOPSIS OF THE EVOLUTION OF ARTIFICIAL INSEMINATION OF HONEY BEE QUEENS

The first attempt to artificially inseminate queens, of which I am aware, was Francis Huber's painting drone semen on the vaginal orifice with a small brush (Huber, 1814). Huber did not succeed in his efforts.

This first attempt was followed by many others until 1927 when Lloyd R. Watson published his research in the booklet *Controlled Mating of Queenbees* and initiated modern efforts to bring the mating of queens under complete control. There is no doubt that some of the workers prior to Watson attained minor successes occasionally because we know now that when semen is introduced into the vagina some sperm reach the spermatheca, and several experimenters had devised techniques similar to Watson's.

It was Watson, however, who refined the instruments and technique and proved that artificial insemination of queens is possible, and many of the queens he treated received partial inseminations. Watson laid the foundation of the work to follow. Besides designing and constructing an efficient syringe, he introduced several other features that are part of artificial insemination as it is practiced today, such as use of a microscope and lamp, alignment of syringe and queen similar to that now used, decapitation of the drone to cause ejaculation, and ejection of a minute amount of semen to lubricate the syringe tip before insertion. He inseminated some queens several times and used semen from more than one drone per injection in other cases. He objected to the term "artificial insemination" and substituted "instrumental insemination" for it.

Watson found a disciple in W. J. Nolan of the United States Department of Agriculture, Division of Bee Culture. Nolan began using the Watson method soon after its announcement (Nolan, 1932), and quickly began to modify the technique and instruments. Nolan's (1937a) efforts culminated in the basic design of the Roberts-Mackensen and Mackensen instruments widely in use today. He also converted a mechanical pencil to a sturdy and efficient syringe that was also used by others until Mackensen developed the syringe that is now part of nearly all instruments of artificial insemination. He used a tube to hold the queen and originated the backup tube to assist in getting the queen into the holding tube.

Harry H. Laidlaw as an undergraduate student at Louisiana State University 1929-33, as a graduate student 1933-34, and as a part-time employee of the United States Department of Agriculture, Southern States Bee Culture Field Laboratory was encouraged to continue his efforts to improve the insemination of queens by direct copulation of queen and drone which had given partial insemination since 1923.

Nolan's queen holding tube was used instead of holding the queen in the hand as had been done previously. During these endeavors, he found that semen was not being deposited in the lateral oviducts as occurs in natural mating (Bishop, 1920). Dissections were made that revealed a structure in the vagina that prevented the forward flow of semen past it. This structure had been described and named the "Ventil-wulst" or valvefold by Ernst von Bresslau in 1906 but had been over-looked by all workers except Bishop (1920) who attached no importance to it.

That discovery led Laidlaw to concentrate on overcoming this ob-stacle. One problem was that the base of the sting overlies the vaginal orifice when the sting chamber is opened, whether with forceps, Nolan's hooks, or with the first "U"-shaped clip Laidlaw used to replace forceps, and interferes with injection of semen into the reproductive tract by any method. To overcome this, Laidlaw constructed a wire clip having a projection at the center of the bend of the "U" which pushed the sting base dorsally to expose the vaginal orifice when the clip was inserted into the sting chamber thus giving access to the vagina. The valvefold still prevented introduction of semen into the oviducts. In the final experiment Laidlaw pushed the valvefold ventrally with a wire loop and semen was squeezed from the bulb of the partially everted penis past the valvefold into the oviducts (Laidlaw, 1934).

The queen's struggling, even when in the kind of tube used by Nolan, made all operations difficult and sometimes resulted in injury to the queen. To avoid these problems, Laidlaw anaesthetized the queens with carbon dioxide or other chemicals.

Later, as a graduate student at the University of Wisconsin, Laidlaw (1939, 1944) abandoned direct copulation as being impractical because getting semen past the valvefold was too difficult. He adopted Nolan's syringe but tapered the tip toward the end so that during injection of semen the vaginal orifice would fit tightly around the tip to prevent leakage of semen. He also used Nolan's hooks to hold the abdomen steady, but the dorsal hook was fashioned to fit between the lancets at the base of the sting so the sting could be pulled from over the vaginal orifice as had been done with the clip. The valvefold was pushed ven-trally in the vagina with a probe and the syringe was inserted into the vagina dorsal to the probe. These innovations made possible the in-jection of semen into the lateral oviducts with little difficulty.

Otto Mackensen (1948) employed as geneticist at the Southern States Bee Culture Field Laboratory began using artificial insemination early in his work at the Laboratory. He modified Nolan's instrument and adopted the sting hook and later the use of carbon dioxide to an-aesthetize the queen during insemination. He made a study of the quan-tity of semen needed to be injected to effect a normal or near normal insemination and determined the effect of multiple injections and the age of the queen best suited for insemination. With the aid of a haemocytometer he counted the sperm in the spermatheca of naturally and artificially inseminated queens. Using the same method he determined the number of sperm produced by drones, and the age at which drones are mature. A similar and confirming study was made by Woyke (1963) who found that the injection of 8 cu mm of semen at one time to be adequate for a normal or near normal insemination. Mackensen at first took semen from the bulb of the partially everted penis of ejaculated drones as was done by Watson and others, but he found that semen released by fully everted drones was equally as good for insemination. Mackensen (1947) also found that queens, including virgins, exposed twice to carbon dioxide at an interval of at least a day laid nearly as soon as naturally mated queens. This discovery solved a vexous problem of delayed oviposition by artificially inseminated queens.

For some years after Watson's announcement of his method, it was still believed by most beekeepers and bee scientists that queens mate naturally with one drone. The better inseminations resulting from using semen from several drones, or from repeated injections of semen, were explained by there being some flaw in the technique of artificial insemination, such as injury to the queen, queens not in physiological condition to receive sperm into the spermatheca, immature or sterile drones, or for other reasons. Multiple mating flights of queens had been reported since Huber observed this in 1789 but these were considered to be abnormal.

In 1932 Nolan reviewed literature which cited observations that some queens take more than one mating flight. In 1940 Oertel verified that some queens take more than one mating flight, and mentions that Mackensen and Roberts in one experiment had nearly half of the queens mate more than once. Roberts (1944) reported half of 110 queens under test mated more than once. Tryasko (1951) dissected queens returning from the mating flight and found that in all cases the paired oviducts were filled with semen, though no mucus, and had much more semen than one drone could supply. It was estimated that the queens had mated with 4 to 5 drones during the one mating flight.

These observations and researches showed that artificial insemination is about as efficient as natural mating in filling the spermatheca with sperm. We do not yet know why such an excess of semen must be deposited in the oviducts by either artificial insemination or natural mating, nor why artificially inseminated queens are slower than naturally mated queens in beginning oviposition.

William C. Roberts (1947) while a graduate student at the University of Wisconsin and a part-time employee of the United States Department of Agriculture North Central States Bee Culture Laboratory designed and constructed a plastic syringe having a removable plastic tip. This was a good syringe, its only real fault was the wear of the bore of the syringe tip by the wire plunger. It was the forerunner of Mackensen's syringe.

Mackensen (1948) avoided this problem of wear by designing a new syringe that incorporated a liquid plunger in a removable plastic tip. The liquid plunger was controlled by a rubber diaphragm in the base of the tip. The syringe was later improved (1954) and the improved model is universally used without further modification except that some workers, though using Mackensen's syringe mechanism, prefer a different tip. Mackensen and Tucker (1970) published descriptions, photographs, and construction drawings of Mackensen's improved instrument and syringe. This instrument is quite easy to use if properly adjusted.

Laidlaw had the prototype of his present instrument made at Wisconsin and used it in his research there. The present version was designed and constructed in California and except for a change in the anaesthetization chamber has not been modified. When one injection of 5 cubic millimeters or more of semen began to be used instead of smaller injections, Laidlaw (1955) experienced excessive losses of inseminated queens because of contamination of semen with normally nonpathogenic bacteria that multiplied rapidly in the semen in the oviducts. He instituted strict sanitary procedures and added an antibiotic to the plunger fluid which greatly reduced losses of queens due to infections.

Tap water, distilled water, and various physiological salt solutions have been used as the liquid plunger of the Mackensen syringe. The sperm would not survive long in any of these solutions, and though usually very little of the solution was injected into the queen with the semen, it was believed that the use of a fluid in which sperm would survive well would enhance the movement of the sperm into the spermatheca. Camargo (1972; 1975) found that coconut water, bacteriologically filtered and adjusted to pH 7 to 7.3 when mixed with semen in a ratio 1 part coconut water and 2 parts semen increased by several times the number of sperm reaching the spermatheca, and recommended coconut water as the liquid plunger. Ruttner (1975) recommended another fluid in which semen survived well. Plunger fluids that are compatible with sperm or which stimulate them constitute an important refinement of instrumental insemination.

The valvefold has been pushed ventrally with a probe, while the syringe is inserted into the vagina, by practically all workers since the procedure was published by Laidlaw in 1944. Vesely (1965) found that the valvefold could be pushed ventrally with the end of the syringe tip instead of with a probe. In 1973 Laidlaw (unpublished) found that the valvefold can be bypassed easily by the syringe if the anterior sting chamber membrane is stretched by pulling the base of the sting dorsally almost as far as it will go without injury to the queen. This pulls the vaginal orifice to the tip of the valvefold or dorsal to it and the syringe is inserted into the open vagina. Proper alignment of the syringe and queen is necessary, and insertion of the syringe must be precise.

Many variations of insemination techniques and instruments have been devised by workers in several countries. Many of the instruments were based on Mackensen's original instrument or the Roberts-Mackensen model; others were not. Ruttner, Schneider, and Fresnaye (1974) designed a "standard" instrument based on Mackensen's original apparatus and encompassing the better innovations originating with themselves and with others, and they published detailed construction drawings (Ruttner, 1975, 1976). Operation of this instrument is similar to that of the present Mackensen instrument except that the movement of the syringe is controlled by screws and racks and pinions. This instrument is the latest in the evolution of artificial insemination instruments.

A. LAIDLAW INSTRUMENT

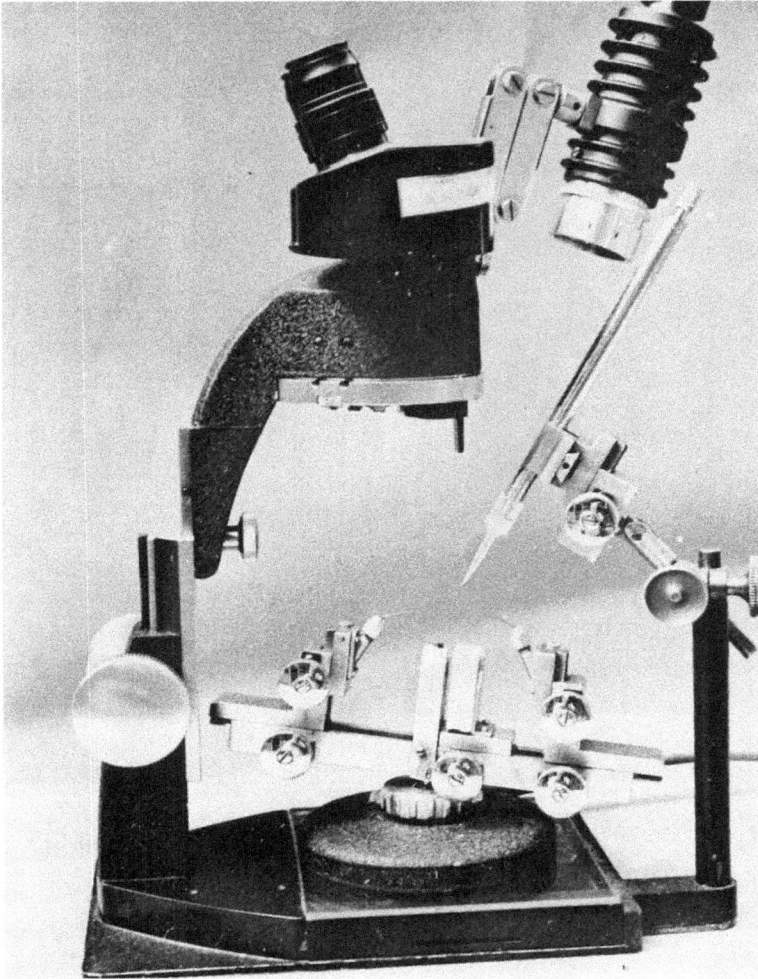

1. LAIDLAW INSTRUMENT AND MICROSCOPE

Instrument is shown on stage of stereoscopic dissecting microscope which is equipped with a 1X objective and 12X or 15X occulars. Lamp has heat filter. Mackensen syringe is used. Queen is held in anaesthetization chamber between sponge rubber pads. Sting chamber is opened and sting pulled dorsally by hooks. Syringe is held in syringe manipulator.

2. LEFT SIDE OF INSTRUMENT showing rubber tubing conducting CO_2 to anaesthetization chamber.

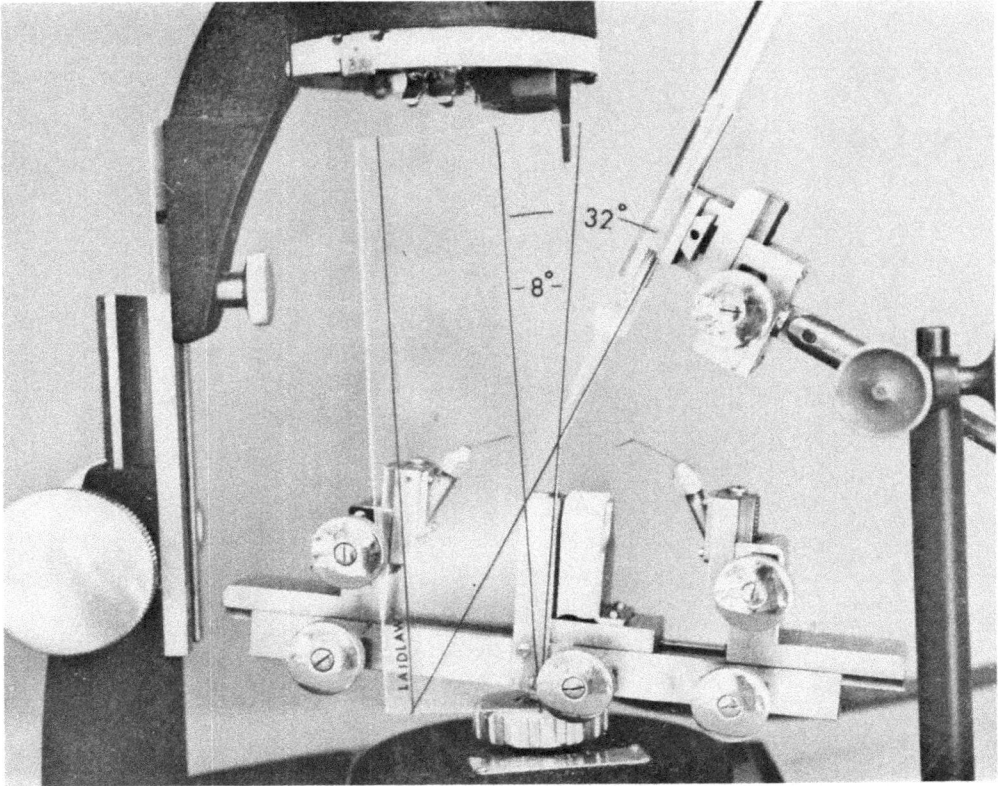

3. LAIDLAW INSTRUMENT ALIGNMENT
Adjustment guide

Adjustment guide is used to properly align queen manipulator and syringe manipulator. Side view. Proper adjustment is necessary for easy and rapid injection of semen.

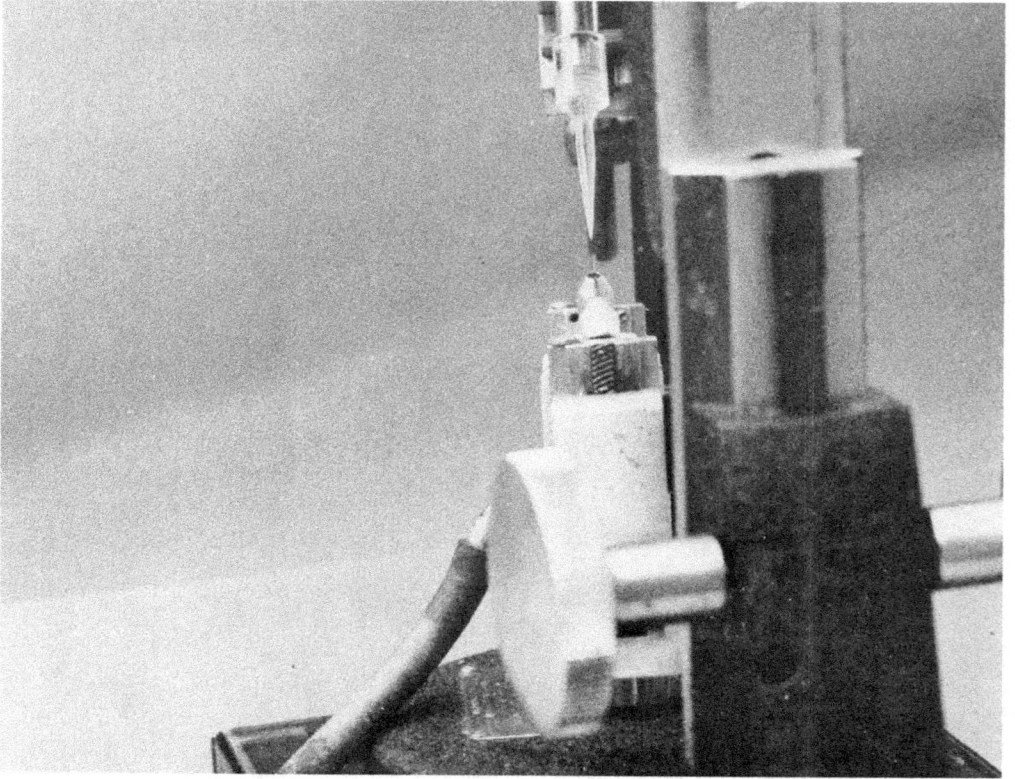

4. LAIDLAW INSTRUMENT ALIGNMENT

Alignment of syringe with queen manipulator — end view.

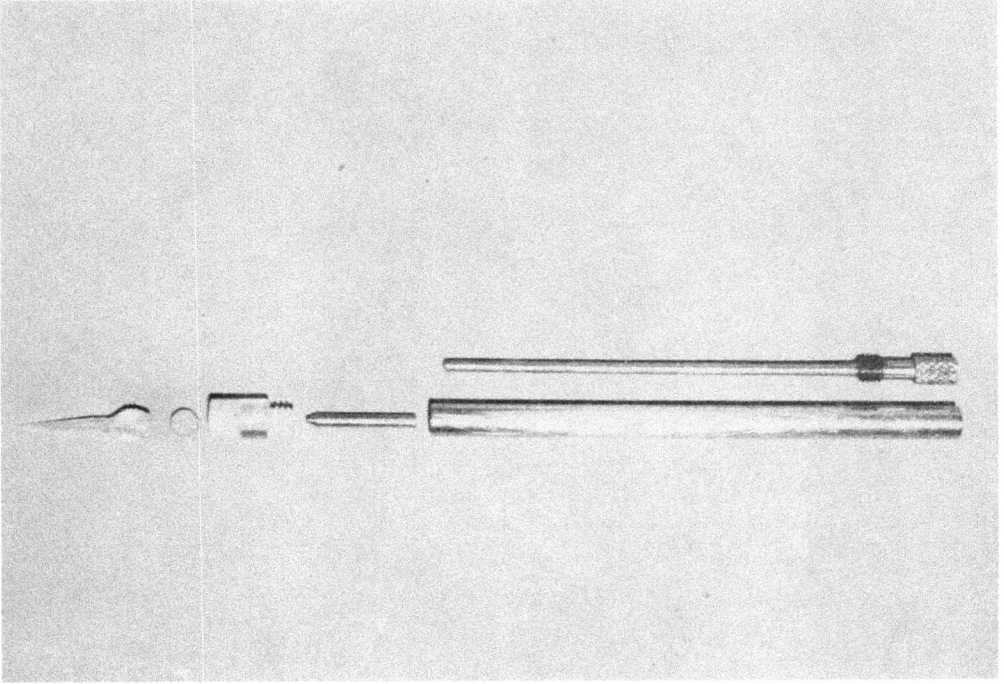

5. MACKENSEN SYRINGE

Exploded view of Mackensen syringe. Plastic tip, rubber diaphragm, plastic adapter, short plunger, barrel, long or screw plunger.

6. MACKENSEN SYRINGE
Base of plastic tip

Base of tip has cone-shaped cavity to receive stretched rubber diaphragm pushed into it by the cone-shaped end of the short plunger.

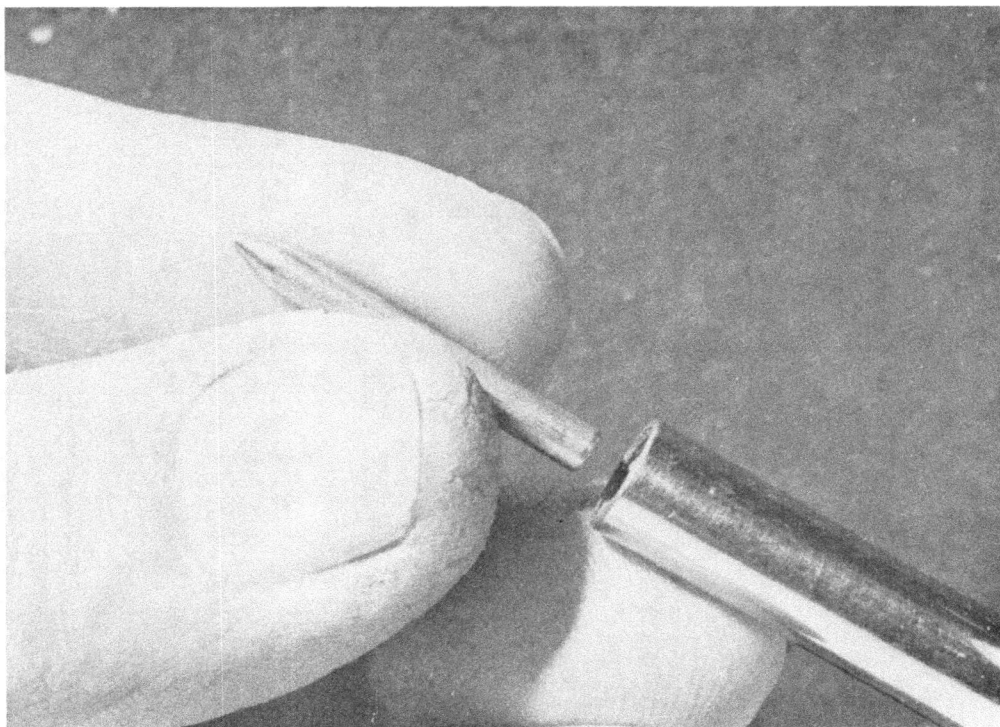

7. ASSEMBLING MACKENSEN SYRINGE

After the long plunger is screwed several turns into the barrel, the short plunger is inserted into the other end of the barrel.

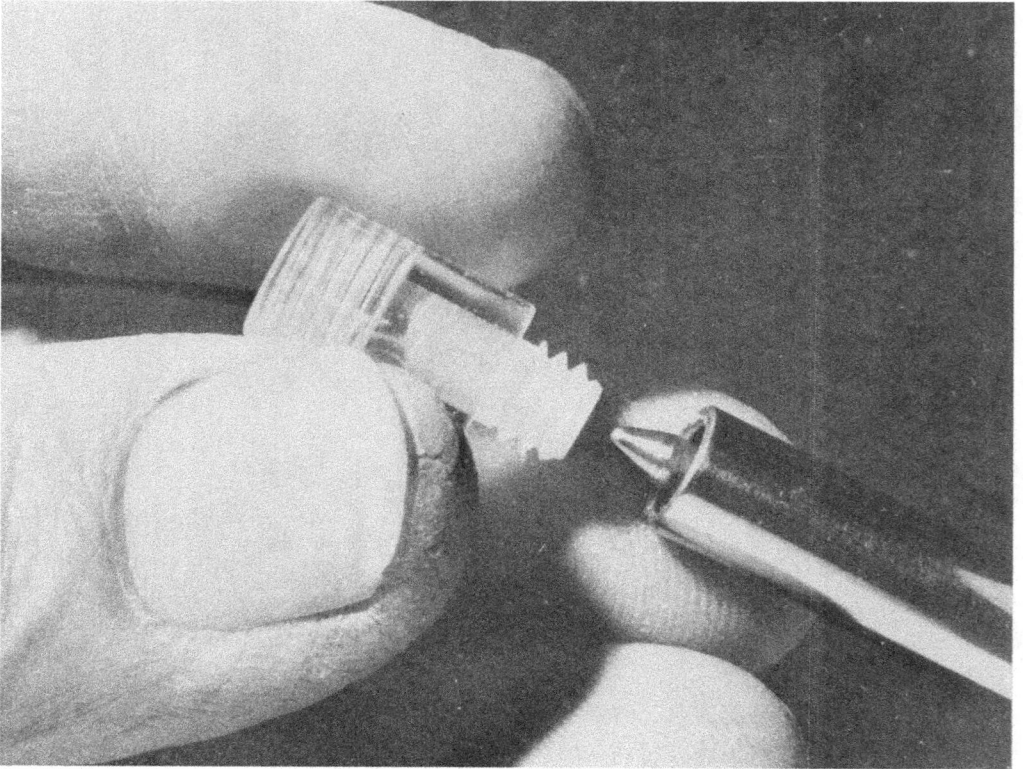

8. ASSEMBLING MACKENSEN SYRINGE

The plastic adapter is now screwed into place firmly but not tightly. Plastic is fragile.

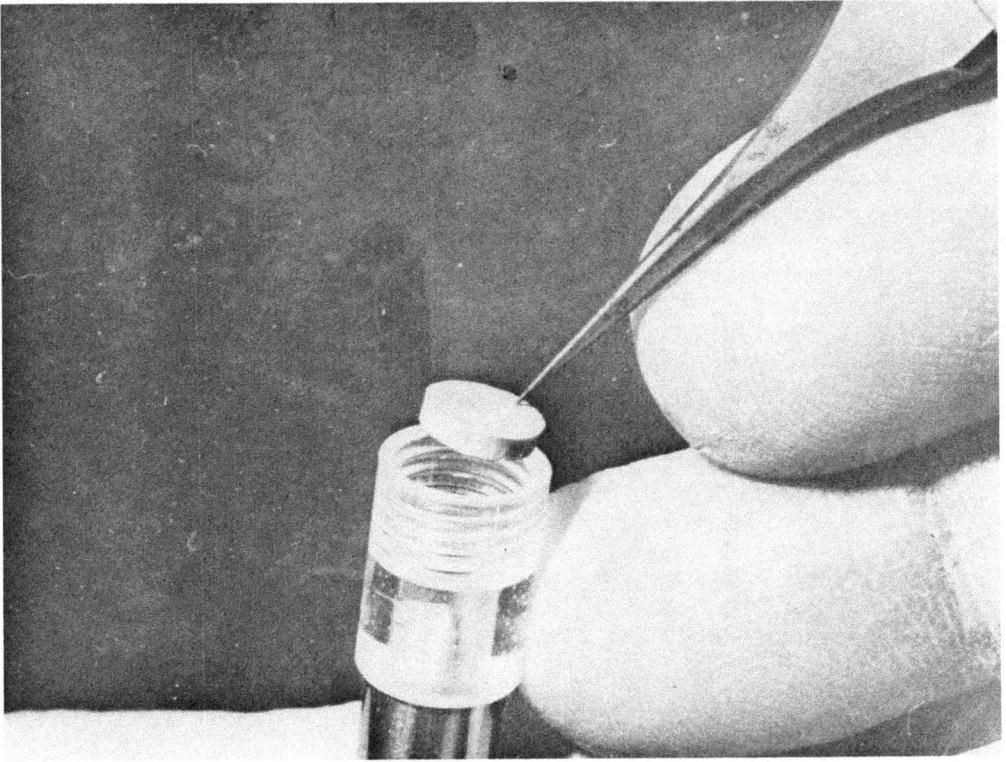

9. ASSEMBLING MACKENSEN SYRINGE

With sterile forceps the diaphragm is placed into the adapter and seated on the inner shoulder.

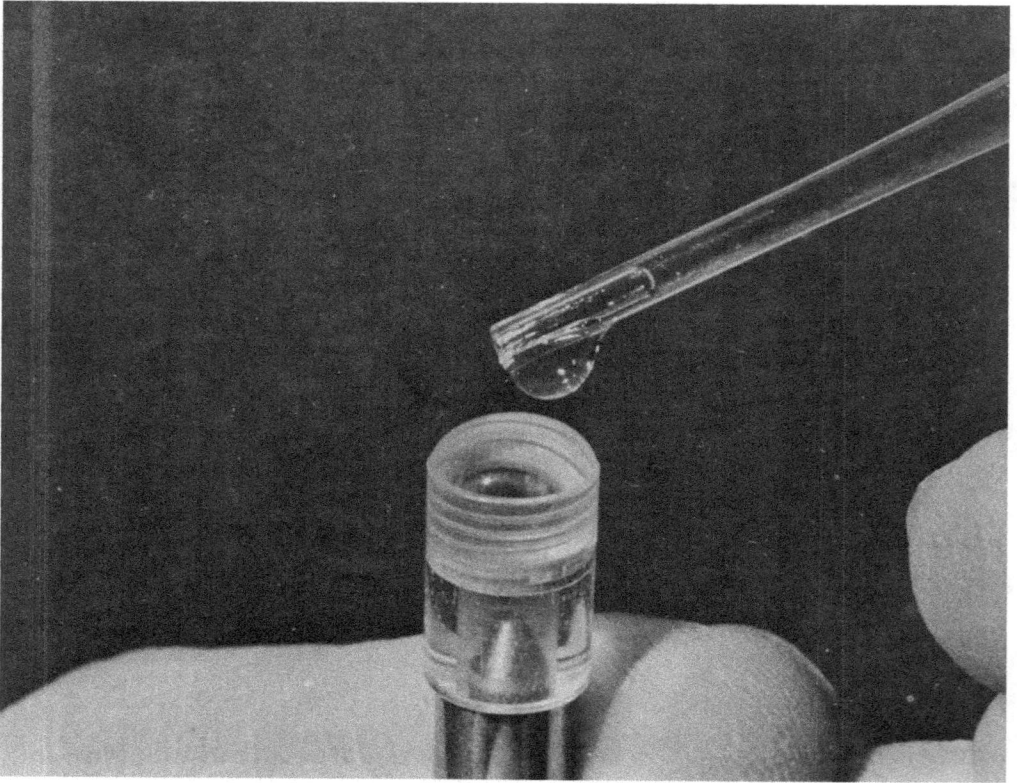

10. ASSEMBLING MACKENSEN SYRINGE

The adapter is now filled with physiological solution containing an antibiotic. One satisfactory solution has the formula 3.5 g NaCl, and 400 cc distilled water, and adjusted to pH 7 to 7.3. Another is coconut water, bacteriologically filtered and adjusted to pH 7 to 7.3, which was introduced by C. A. Camargo (1972). When mixed with semen in the proportion of 1 part coconut water to 2 parts semen, the number of sperm that reach the spermatheca is greatly increased. 30 to 50 cc of either solution is placed in a bottle and dihydrostreptomycin sulfate added to make 0.25% of the total.

Ruttner (1975) recommends a physiological solution made as follows: trisodium-citrate 2-hydrate, 2.43 grams; sodium bicarbonate 0.21 grams; potassium chloride 0.04 grams; sulfanilamide 0.30 grams; d-glucose, 0.30 grams; distilled water 100.00 ml.

Sterilize by heating to 90°C. This solution discolors if overheated.

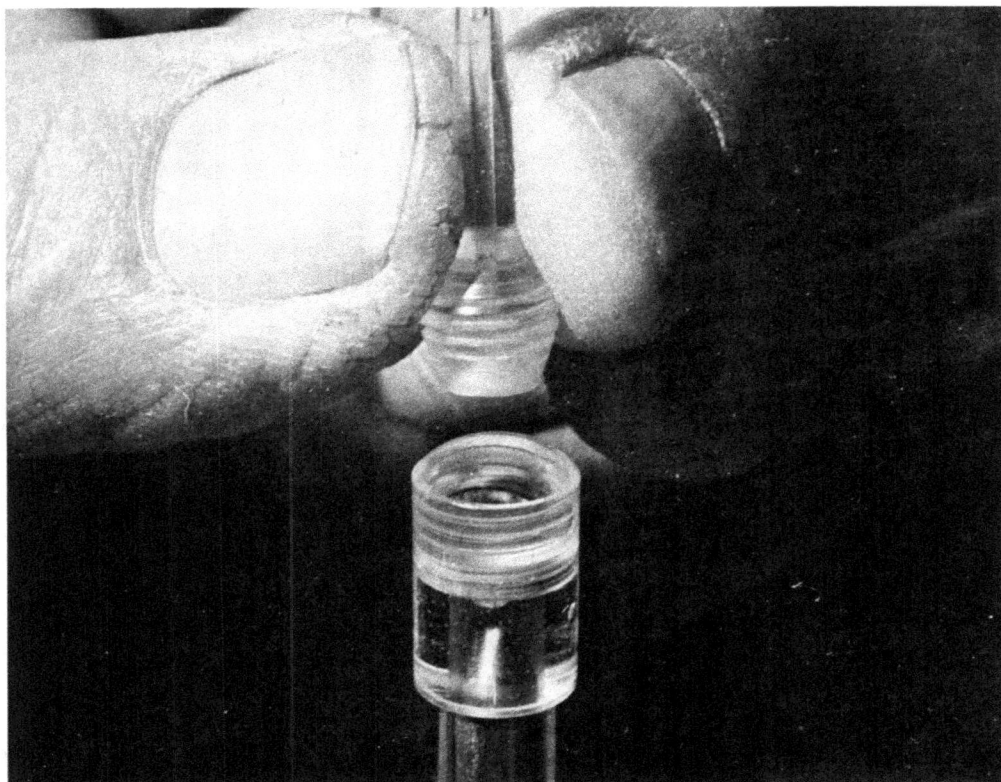

11. ASSEMBLING MACKENSEN SYRINGE

The tip is screwed into the adapter until it seats firmly against the rubber diaphragm.

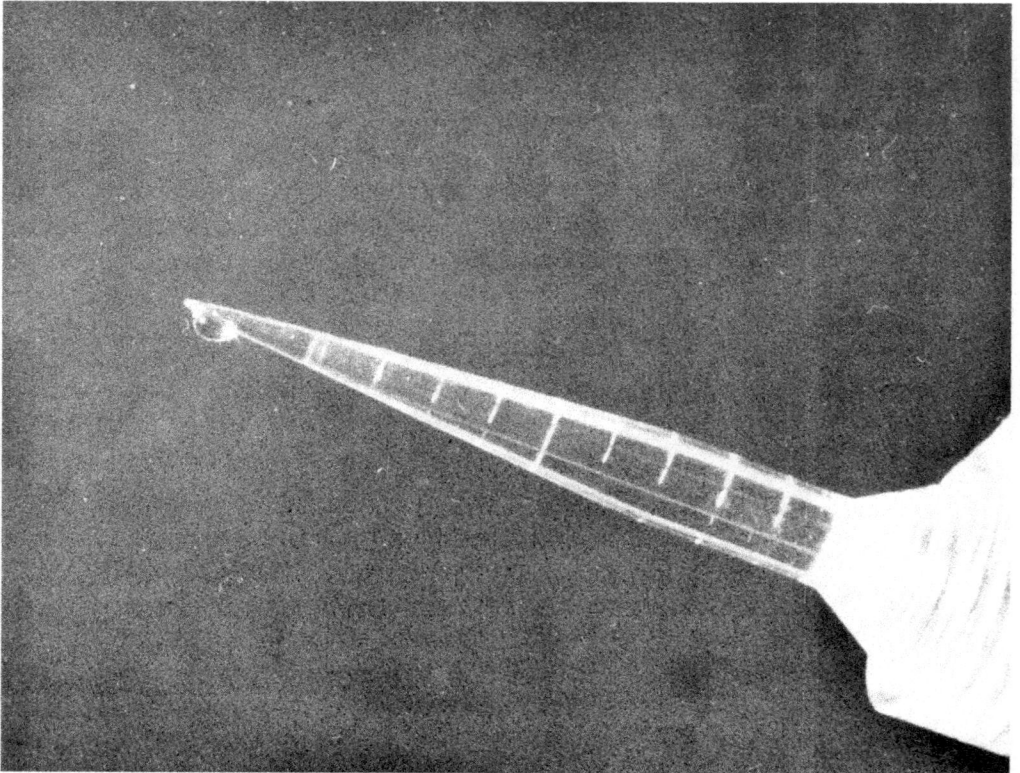

12. ASSEMBLING MACKENSEN SYRINGE

The tip at this point is filled with fluid and part of it must be ejected before semen can be drawn into the tip. The fluid plunger, rather than expandable and compressible air, is necessary for smooth functioning of the syringe because the tip orifice is minute and semen dries rapidly in surrounding air to form a seal across it. Fluid is ejected until upon reversal of the screw plunger the fluid will be withdrawn to the threads of the base of the tip. This provides space for taking up about 10 cubic millimeters (or 10 microliters) of semen, though normally 8 cubic millimeters are injected into the queen at one time. Ejection of too little fluid will prevent filling the tip with 8 cu mm of semen. Ejection of too much fluid may not allow injection of the entire load of semen into queen because the liquid plunger would be too short.

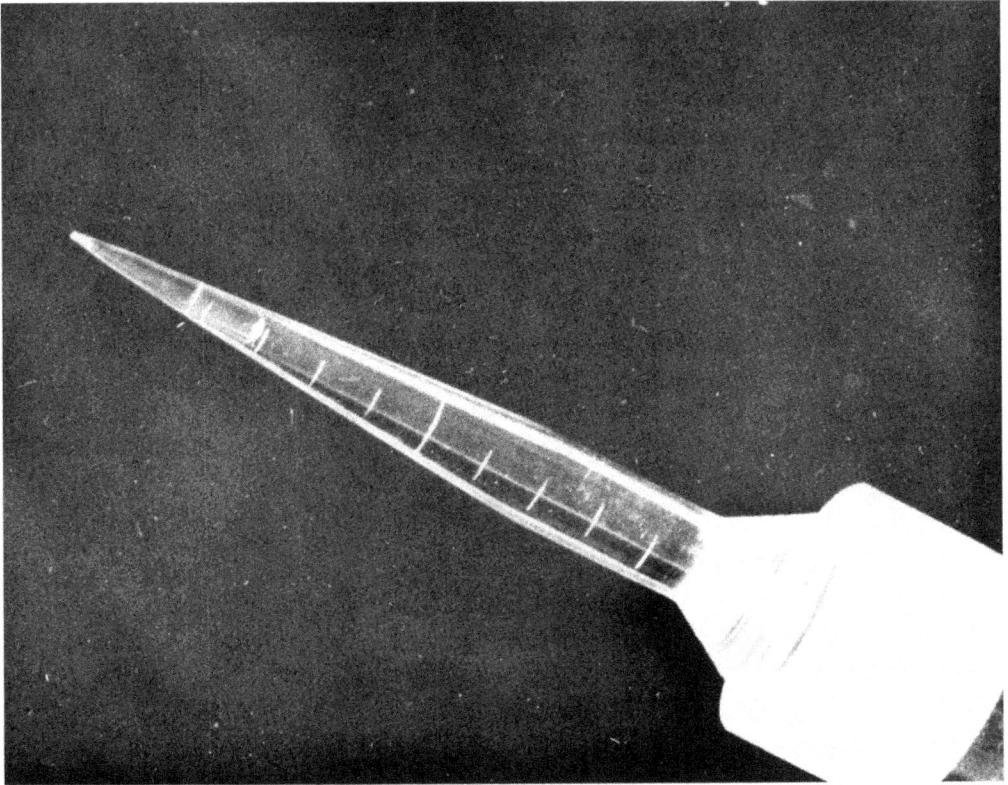

13. ASSEMBLING MACKENSEN SYRINGE

The liquid plunger is moved down the tip until it nearly reaches the distal calibration mark on the tip. This creates an air space between the plunger and the end of the plastic tip which becomes a bubble between the liquid plunger and the semen and prevents admixture of the two.

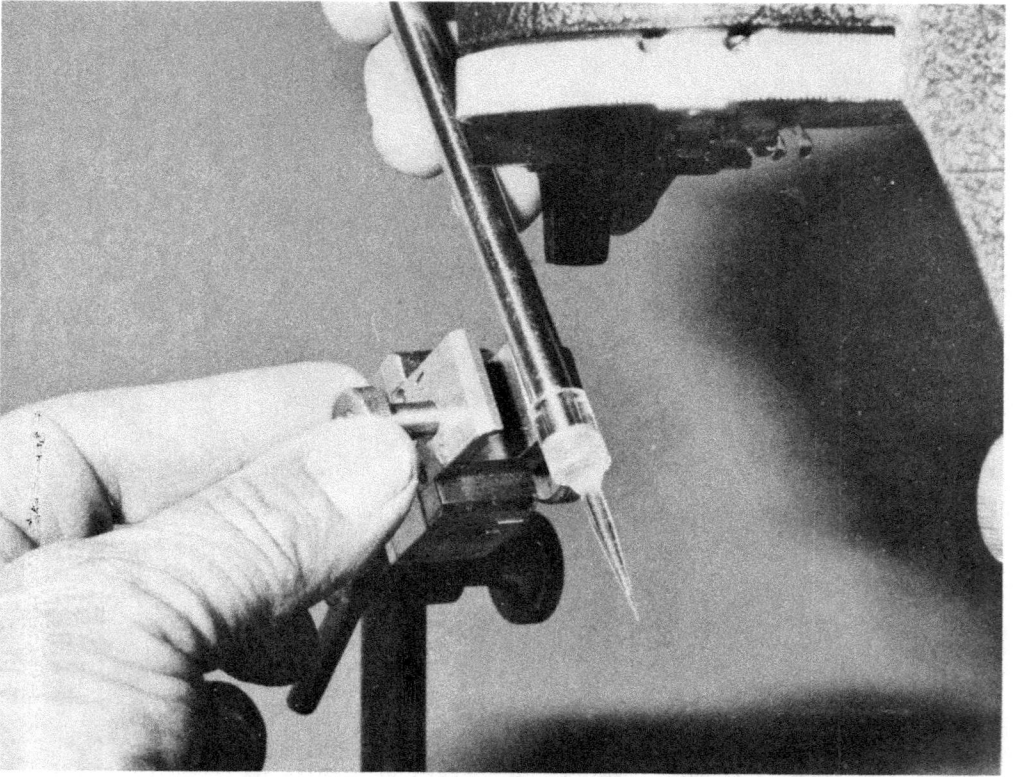

14. FASTENING SYRINGE IN SYRINGE MANIPULATOR

The prepared syringe is slipped into the opened jaws of the syringe manipulator and firmly secured.

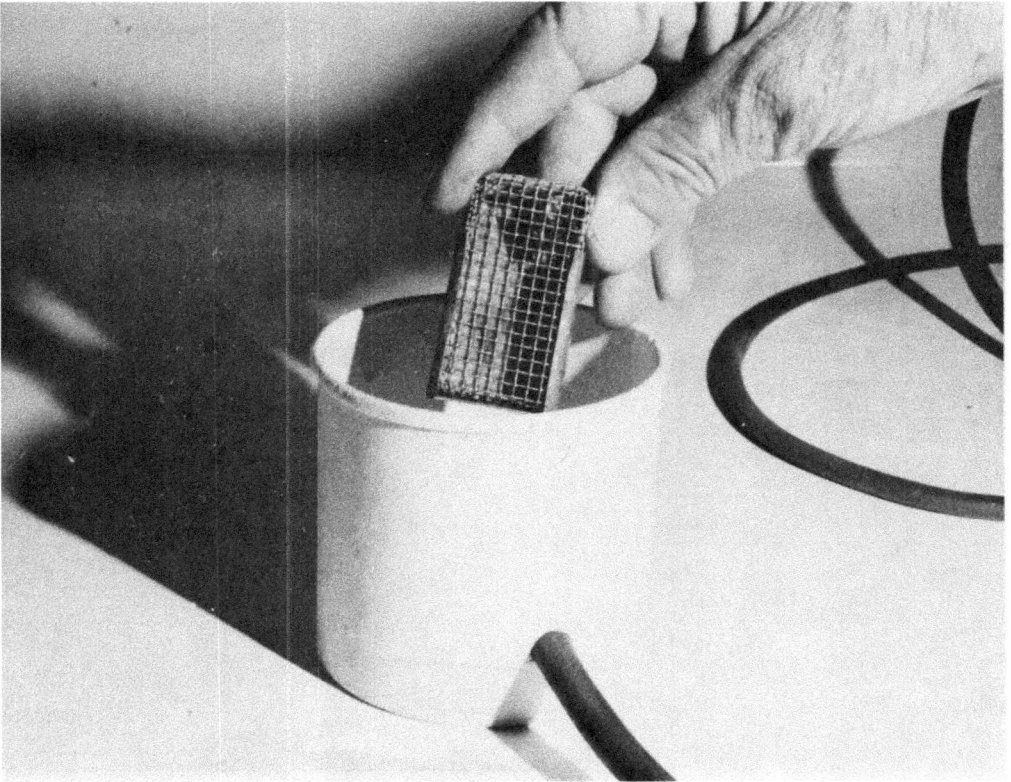

15. PRELIMINARY ANAESTHETIZATION OF QUEEN

Immediately following the preparation of the syringe, the caged virgin is placed in a low container into which a slight continuous flow of carbon dioxide enters near the bottom.

16. OBTAINING SEMEN
Flight box

A few drones (15-20) are released in a flight box having a hole at each end closed with overlapping rubber sheeting through which the drones can be removed. A light above stimulates drone flight. Active drones ejaculate better than sluggish ones but soon tire in the flight box. The drones are easily caught and some final selection can be made for activity, color, or other traits.

17. OBTAINING SEMEN
Ejaculating drone

Very active mature drones may evert the penis and ejaculate semen when they are grasped. Most, however, require other stimulation. The drone should be grasped by the thorax between thumb and forefinger with the ventral side toward the thumb. Now slide the thumb toward the head, straightening it in line with the axis of the thorax, and crush the head, pressing at the same time on the anterior half of the thorax. The thumb nail pressing against the middle of the underside of the thorax will often bring about eversion and ejaculation by drones that do not react otherwise.

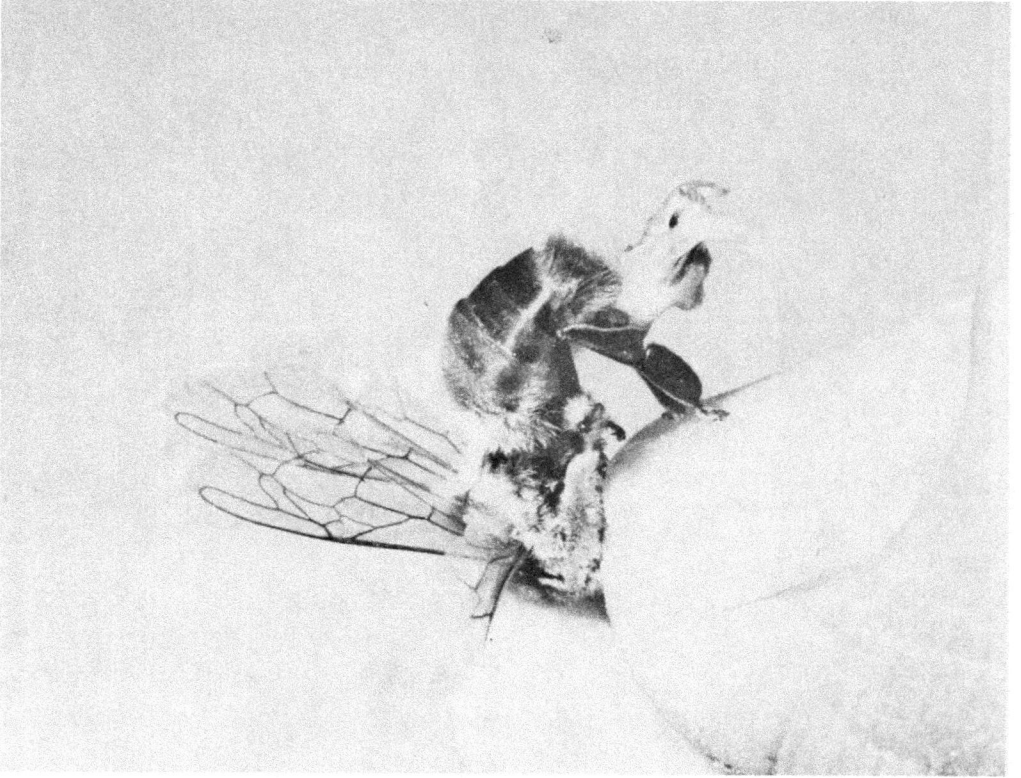

18. OBTAINING SEMEN
Partial eversion

Induced eversion usually stops before completion. If the abdomen contracts and becomes hard, ejaculation of the semen has usually occurred though the semen is contained in the part of the penis still uneverted, in a part called the "bulb," and is yet inaccessible. If the abdomen remains soft at this stage, ejaculation usually has not occurred, and the drone should be discarded.

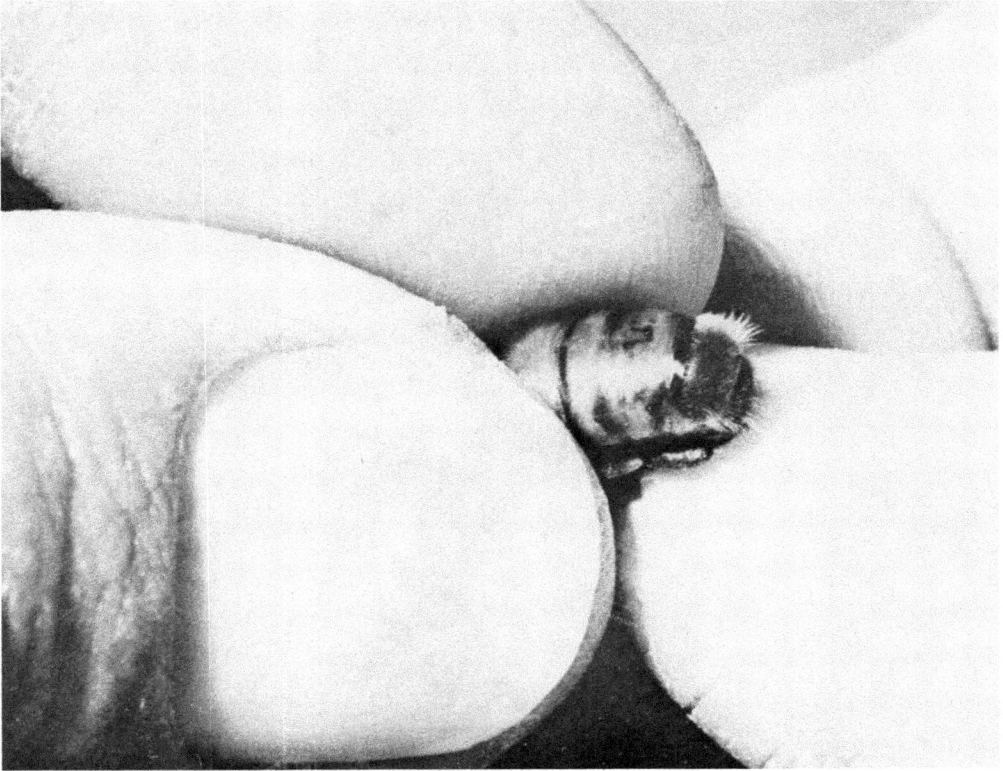

19. OBTAINING SEMEN
Alternative method of ejaculating drones

Some mature drones do not react as just described. Partial eversion and ejaculation in these drones can usually be brought about by pushing the abdominal segments toward the thorax so the blood forces the penis out. But it is essential that the abdominal muscles be **stimulated** to contract; do not crush the abdomen.

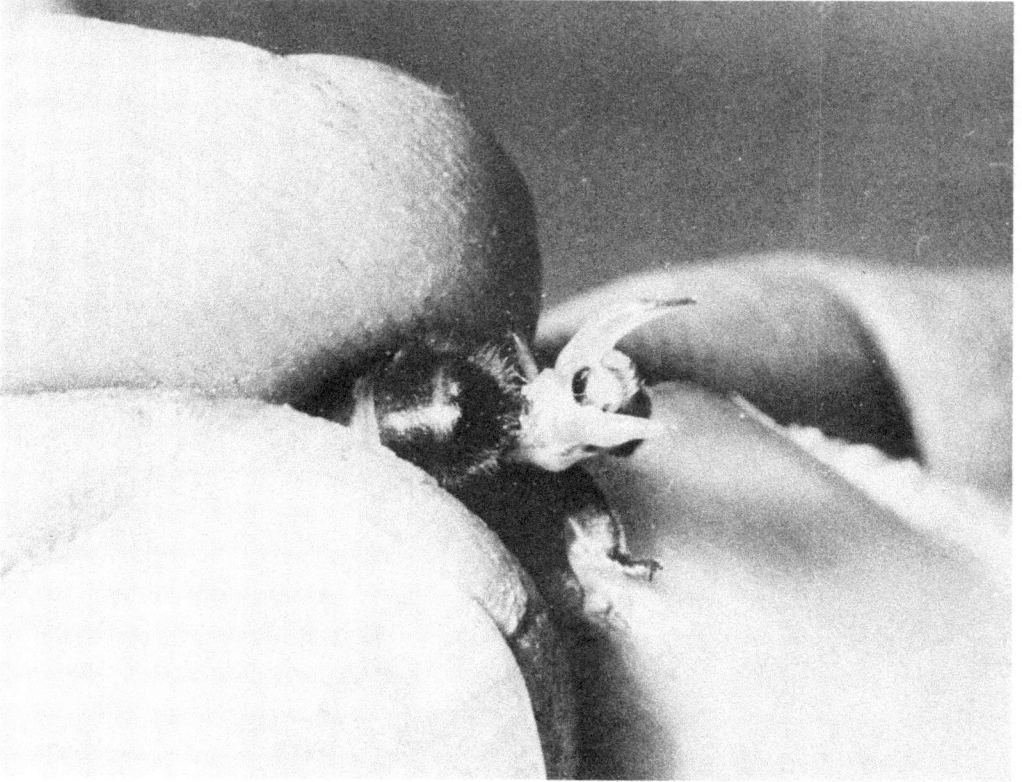

20. OBTAINING SEMEN
Initial eversion by alternative method

Note: Exposure of the drone to chloroform fumes will usually induce partial eversion and ejaculation, as will contact with weak electric current, or exposure to some other chemicals or heat or cold. Drones ejaculated by any method should be used without much delay because the sperm tend to move into the mucus and the semen becomes difficult to separate from the mucus.

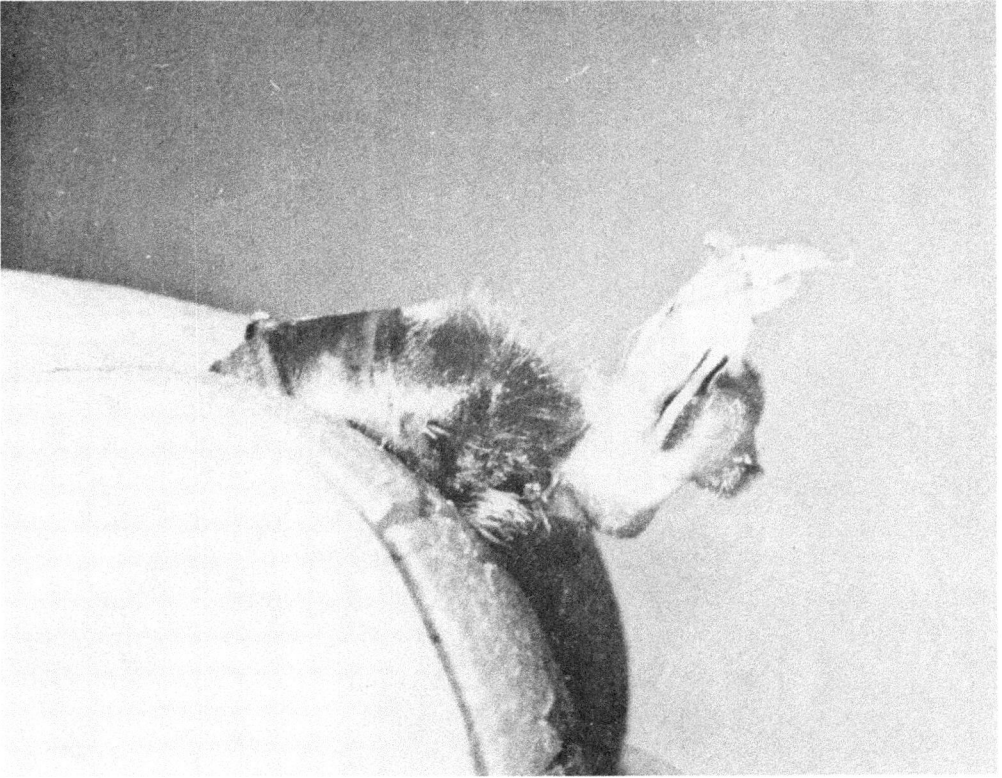

21. OBTAINING SEMEN
Completion of eversion

To release the semen to the exterior of the penis, the eversion must be continued until the semen appears. Usually this can be accomplished by squeezing the drone between the thorax and abdomen either laterally or dorso-ventrally.

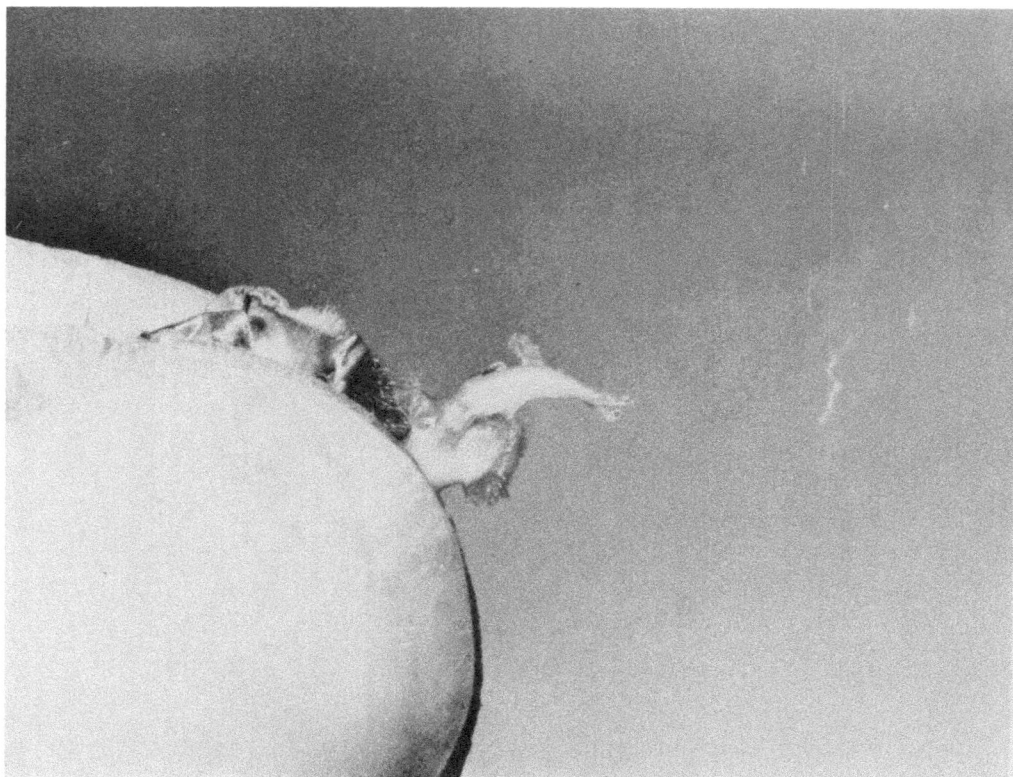

22. OBTAINING SEMEN
Completion of eversion

In some cases it is necessary to apply pressure to the base of the partially everted penis. In this case the base of the penis should be at the end of the thumb and finger and the dorsal side of the abdomen should be parallel and level with the dorsal side of the forefinger. Otherwise, the natural curving of the everting penis may cause the semen to touch the finger and possibly become contaminated. To continue the eversion and release the semen, the wrist is rotated so the thumb and forefinger of the left hand can be brought up alongside the drone's abdomen. The thumb and forefinger are then rolled together toward their tips, progressively mashing the abdomen toward the penis and finally the penis base itself. This gives positive control.

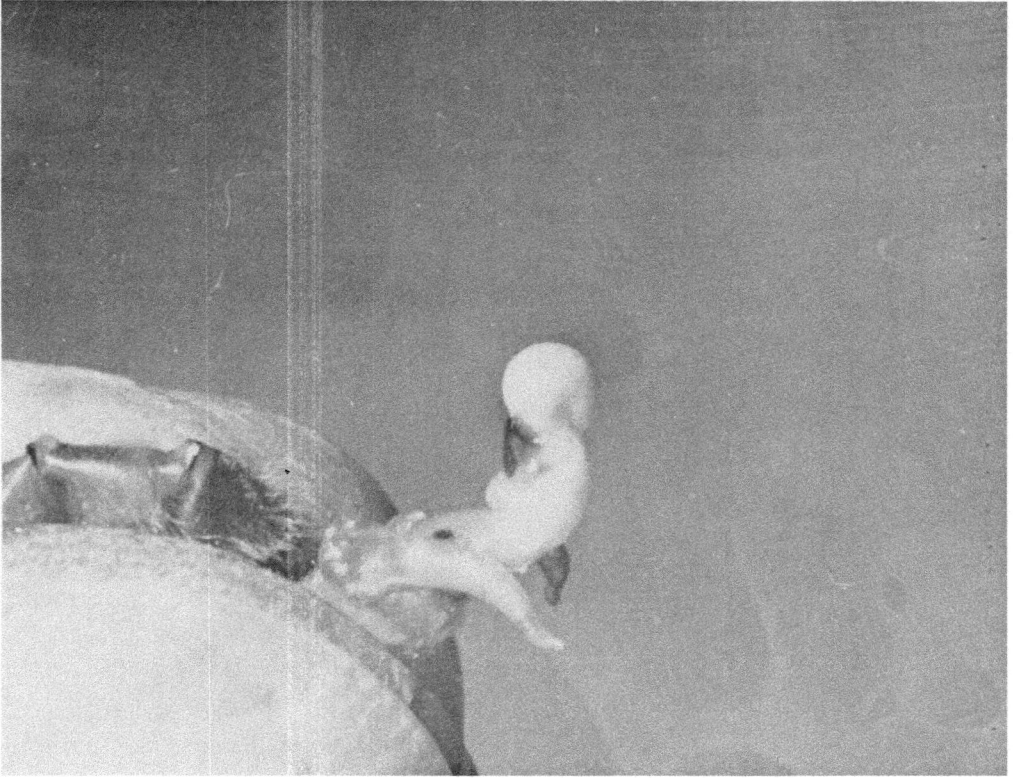

23. OBTAINING SEMEN
Release of semen to the exterior

When semen appears, pressure is relaxed. It is imperative that the semen does not touch any part of the drone or fingers nor become contaminated with fecal material that may flow down the penis to the semen.

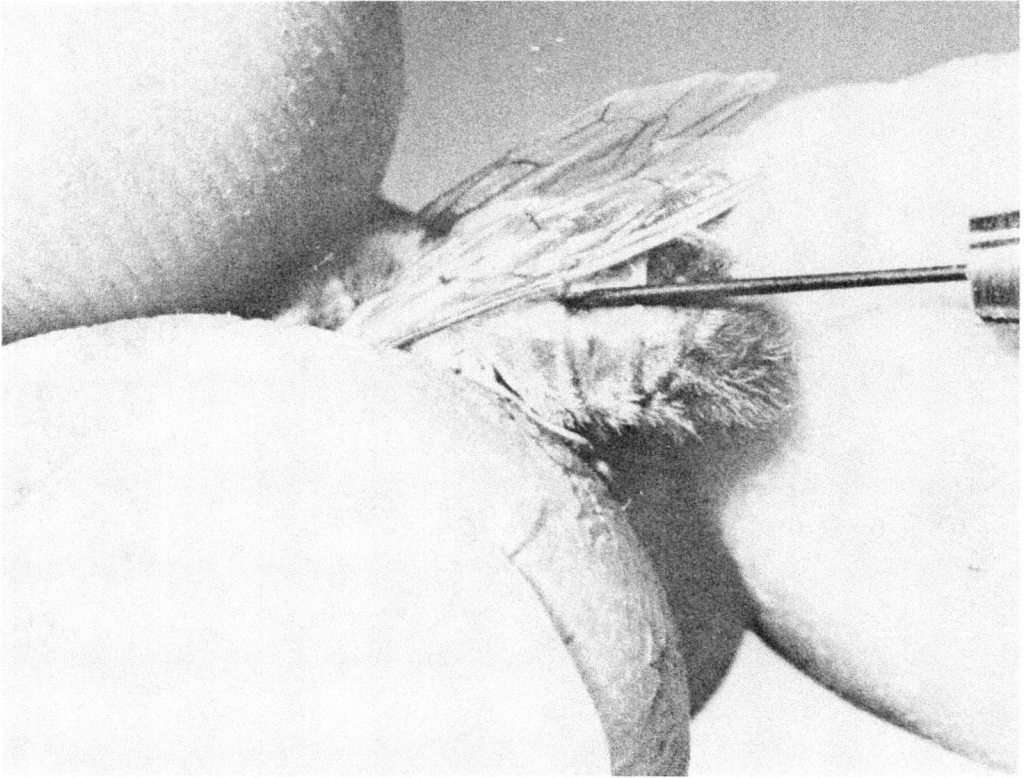

24. OBTAINING SEMEN
Too violent eversion

At times drones of some stocks react to stimulation with explosive eversion and ejaculation. The semen may be propelled some distance away, or the penis may burst and the semen become contaminated with the drone's blood. This can be prevented by piercing the basal region of the abdomen to make a hole for the escape of blood as the abdomen contracts. **Do not pierce the uneverted penis.**

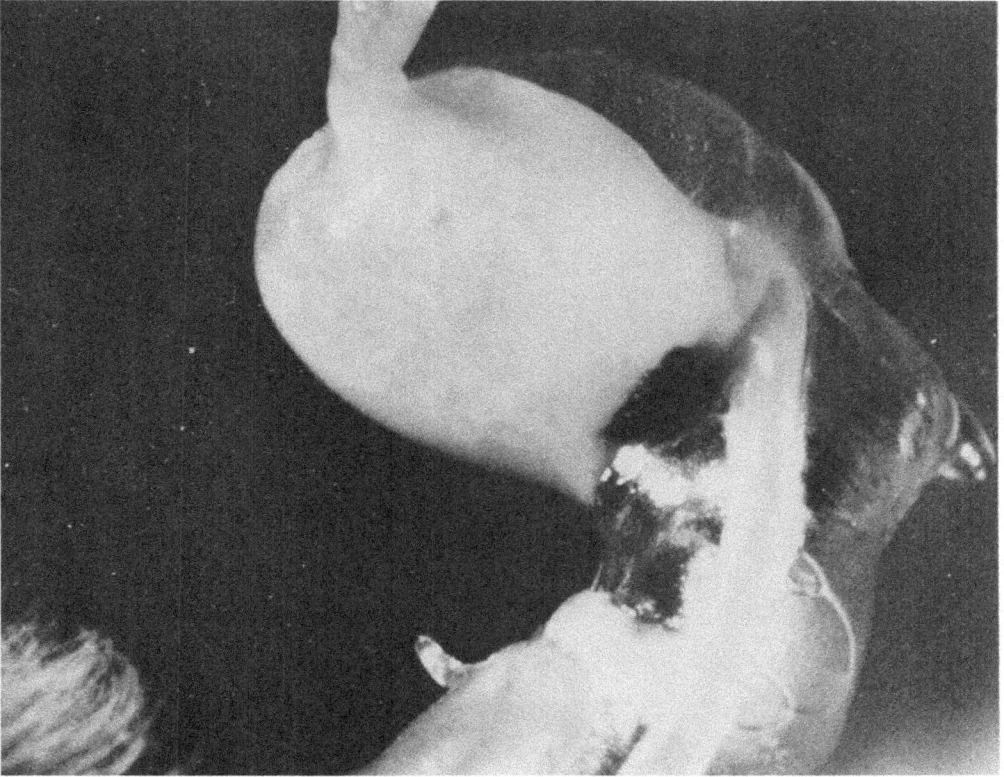

25. TAKING SEMEN INTO THE SYRINGE

The end of the syringe tip is brought into focus in the center of the microscope field and the semen adhering to the penis is barely touched to the end of the syringe tip. Never thrust the tip into the semen because it will pass through the semen into underlying mucus. Turn the long plunger counterclockwise slightly and as the semen flows into the syringe move the penis away from the syringe tip to form a column or cone of semen. Often all of the semen is drawn off of the underlying mucus. If this does not occur, the semen may be skimmed by moving the penis and semen under the syringe bringing all of the semen to the tip. Sometimes semen is on both sides of the penis and it can be brought to the syringe tip by rotating the drone between the thumb and forefinger.

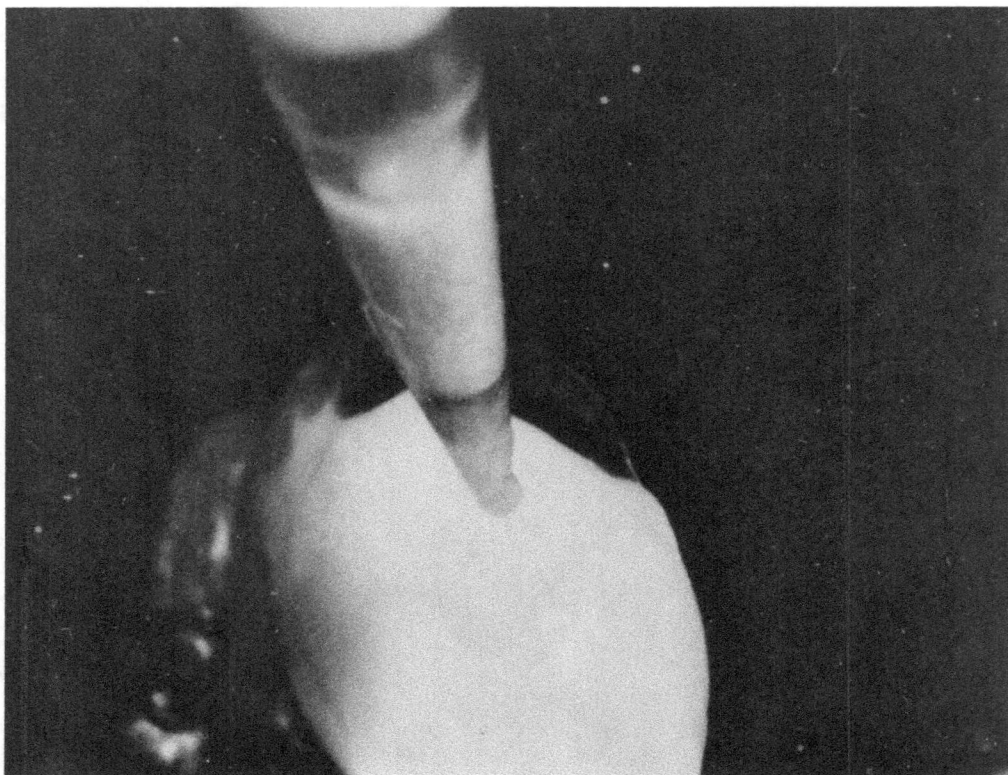

26. TAKING SEMEN INTO THE SYRINGE

If contact of the tip with semen is lost, a small amount of semen should be ejected and this touched to the semen on the penis to regain contact. The same procedure should be used in beginning to take semen from the next drone. Between each drone, withdraw the semen a millimeter or more into the syringe to prevent the formation of a seal at the end of the tip.

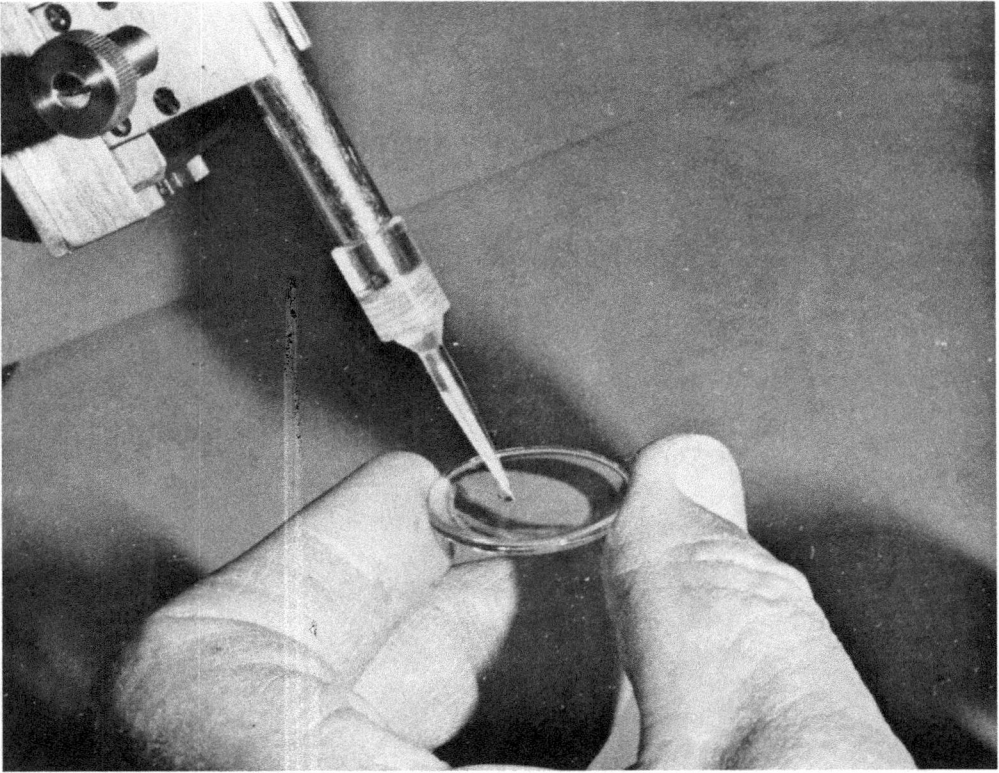

27. TAKING ANTIBIOTIC INTO THE SYRINGE

When the syringe is loaded, about ½ cubic millimeter of the plunger fluid containing an antibiotic is drawn into the tip. This serves to prevent infection, to prevent the formation of a seal at the end of the syringe tip, and as a lubricant when the tip is inserted into the queen.

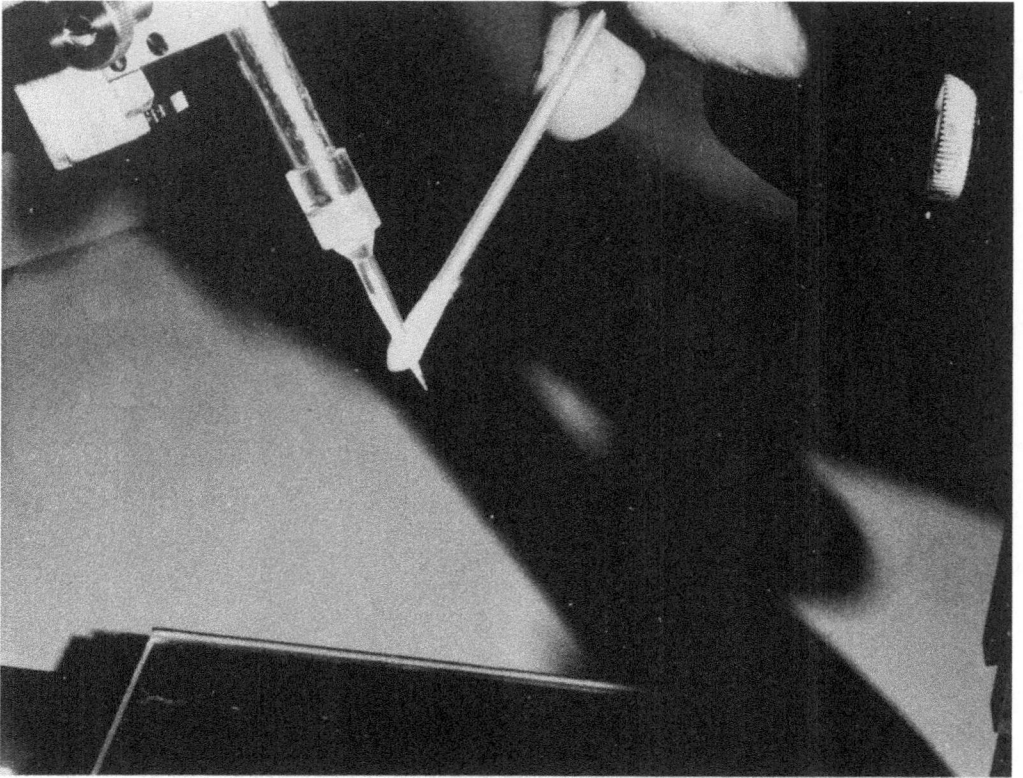

28. WIPING SYRINGE TIP

Any bit of dried semen or mucus on the lower outside sidewalls of the syringe tip will interfere with insertion of the tip into the queen. The tip may be cleaned by wiping with sterile tissue or with a cotton swab dipped in the plunger fluid. After use, the tissue or swab is discarded. Never wipe the tip with fingers after it has been placed in the syringe manipulator, and always wash hands before assembling the syringe or changing tips.

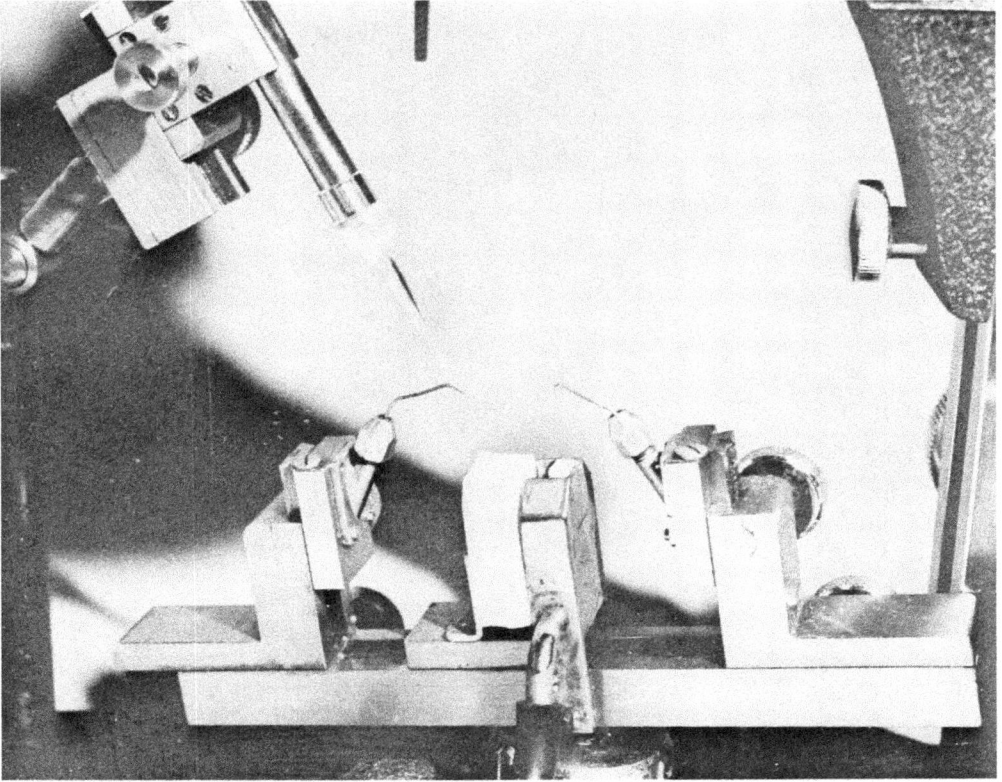

29. RAISING SYRINGE

The filled syringe is now raised as far as possible so the tip will clear
the raised sting hook of the queen manipulator when the manipulator
is put on the microscope stage.

30. PLACING QUEEN ON QUEEN MANIPULATOR

The queen manipulator is set on the table to the side of the microscope and the carbon dioxide is adjusted to a very light flow to the anaesthetization chamber. The anaesthetized queen is now taken from the cage in the anaesthetization container and held between the thumb and forefinger of the right hand with the dorsal side of the queen against the thumb. At this time inspect the queen for injury, malformation, or any other undesirable feature. Discard such queens.

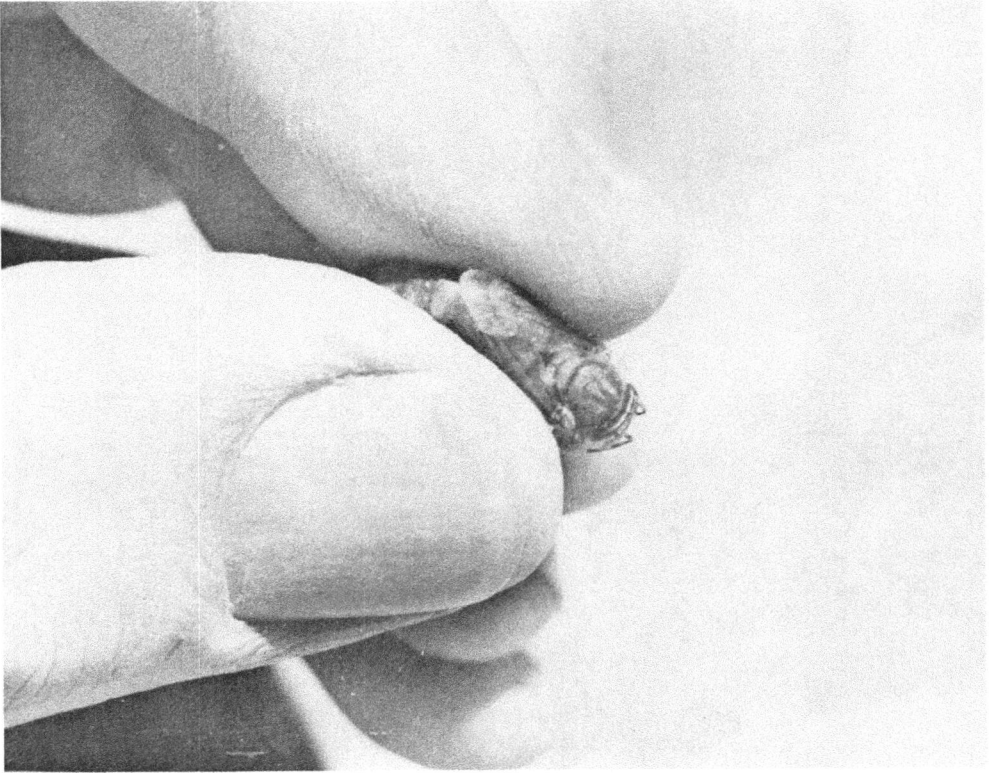

31. PLACING QUEEN ON QUEEN MANIPULATOR

The basal segments of the queen's abdomen are now grasped lightly between the thumb and forefinger of the other hand.

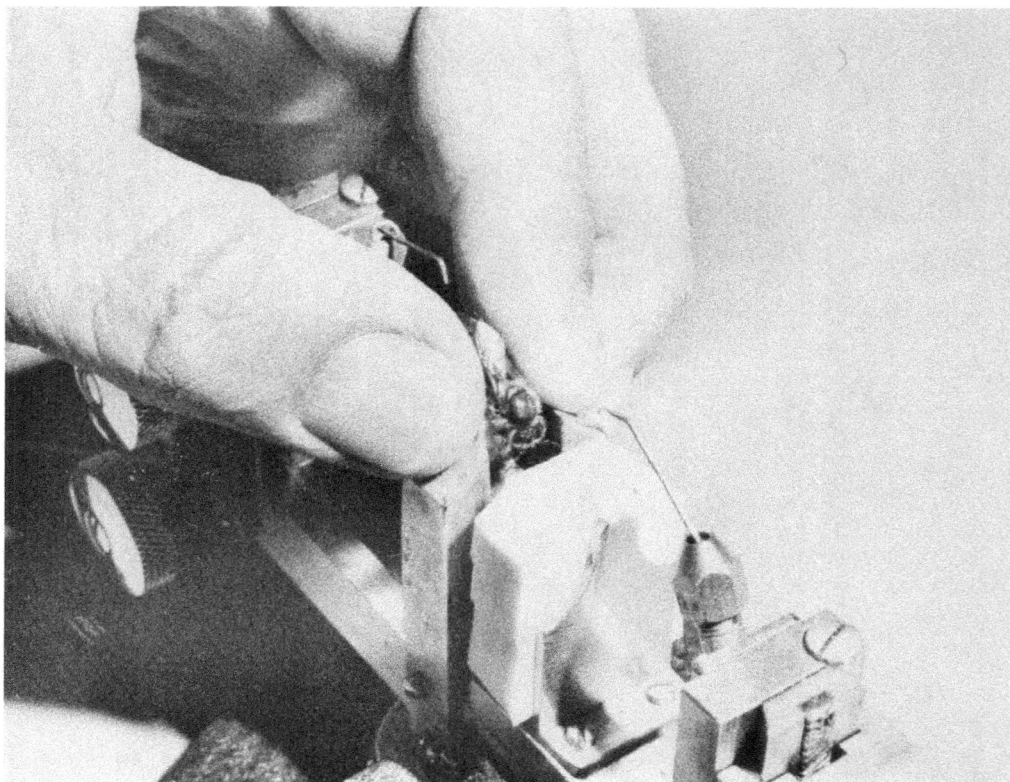

32. PLACING QUEEN ON QUEEN MANIPULATOR

The queen is centered in the anaesthetization chamber of the queen manipulator with the aid of the sloping sides of the chamber. The queen may be rotated between the fingers to properly align her.

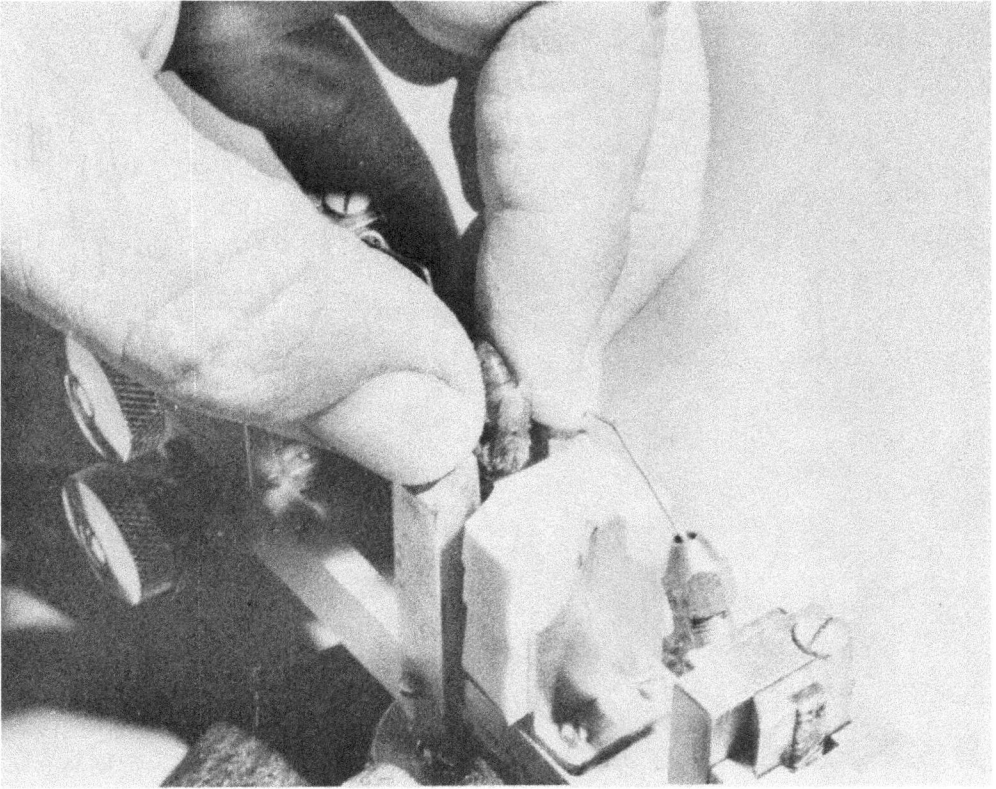

33. PLACING QUEEN ON QUEEN MANIPULATOR

The wrist is now raised which lowers the queen into the chamber, and the movable closing piece faced with fine sponge rubber is brought firmly, but not tightly, against the queen's thorax.

34. PLACING QUEEN ON QUEEN MANIPULATOR

Queen properly fastened in the anaesthetization chamber with abdomen protruding. It is only necessary for the CO_2 to be directed against the anterior thoracic spiracles to bring about complete anaesthetization.

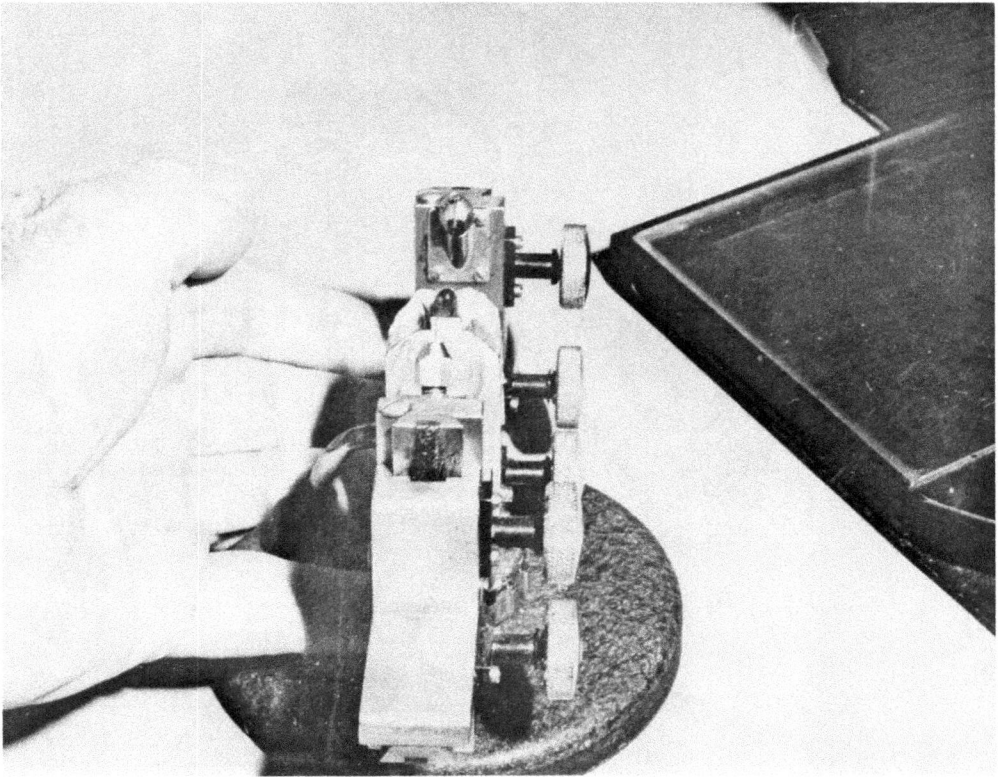

35. MOVING QUEEN MANIPULATOR
TO MICROSCOPE STAGE

The queen manipulator with queen is moved to the microscope stage by slipping fingers between the base and the bar of the manipulator and lifting it onto the stage.

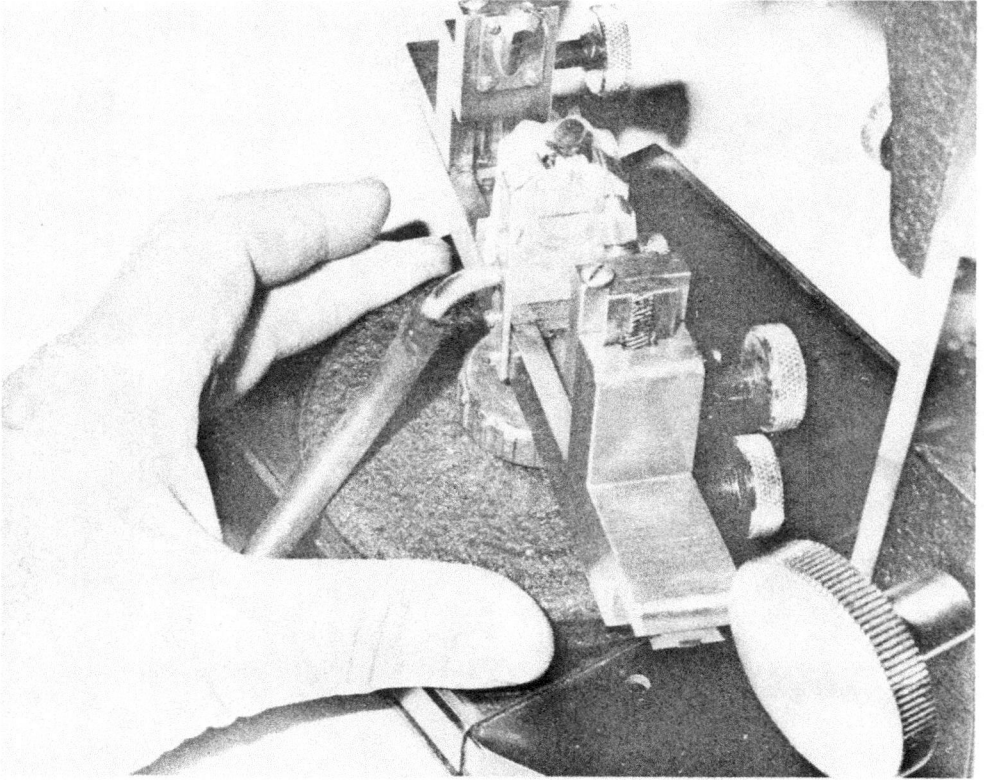

36. MOVING QUEEN MANIPULATOR
ON THE MICROSCOPE STAGE

To properly position the queen in relation to the syringe, the queen manipulator is moved on the stage by grasping the base and steadying the fingers on the stage surface and sliding the base on the stage. The stage must be clean and free of bits of wax or any other substance.

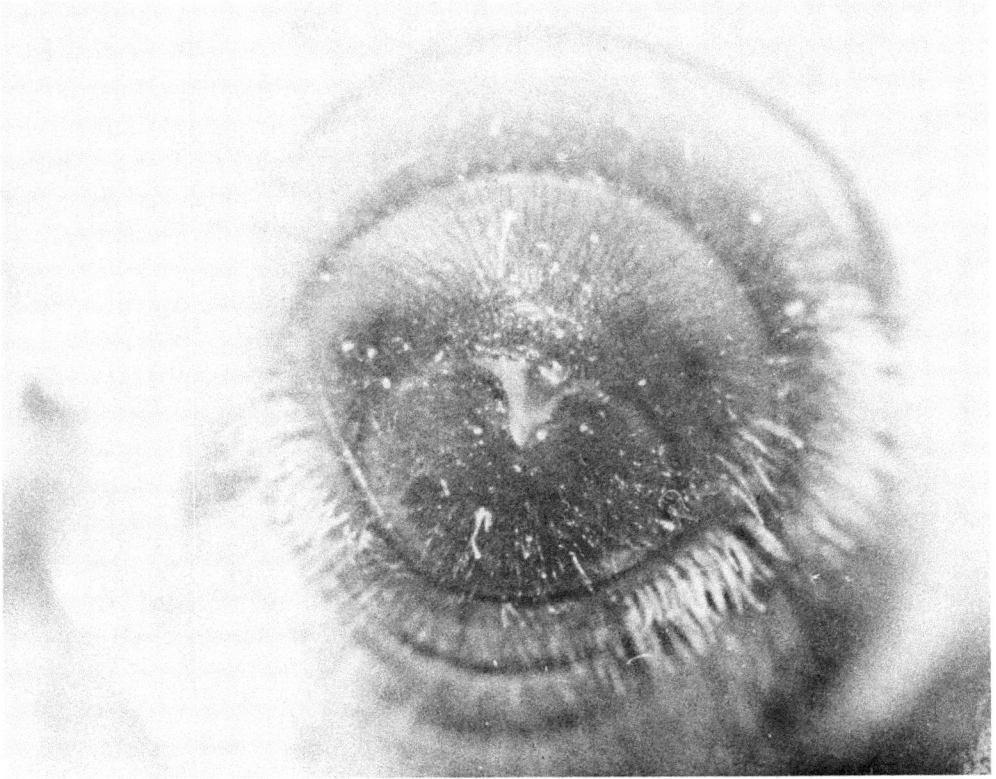

37. QUEEN IN PROPER POSITION ON STAGE

Queen in proper position on stage for opening sting chamber.

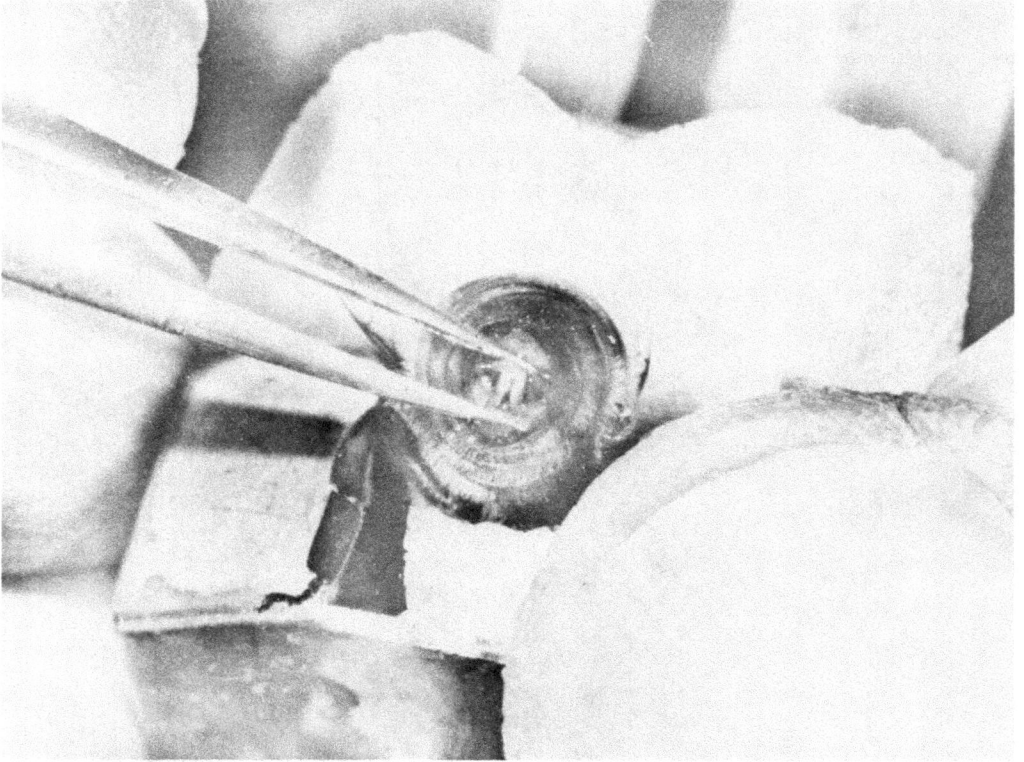

38. OPENING STING CHAMBER

The thumb of the right hand is brought against the corner of the anaesthetization chamber and against the abdomen of the queen. Fine forceps held in the left hand are inserted into the sting chamber. The left hand is steadied by the 2nd and 3rd finger tips resting lightly against the side of the anaesthetization chamber.

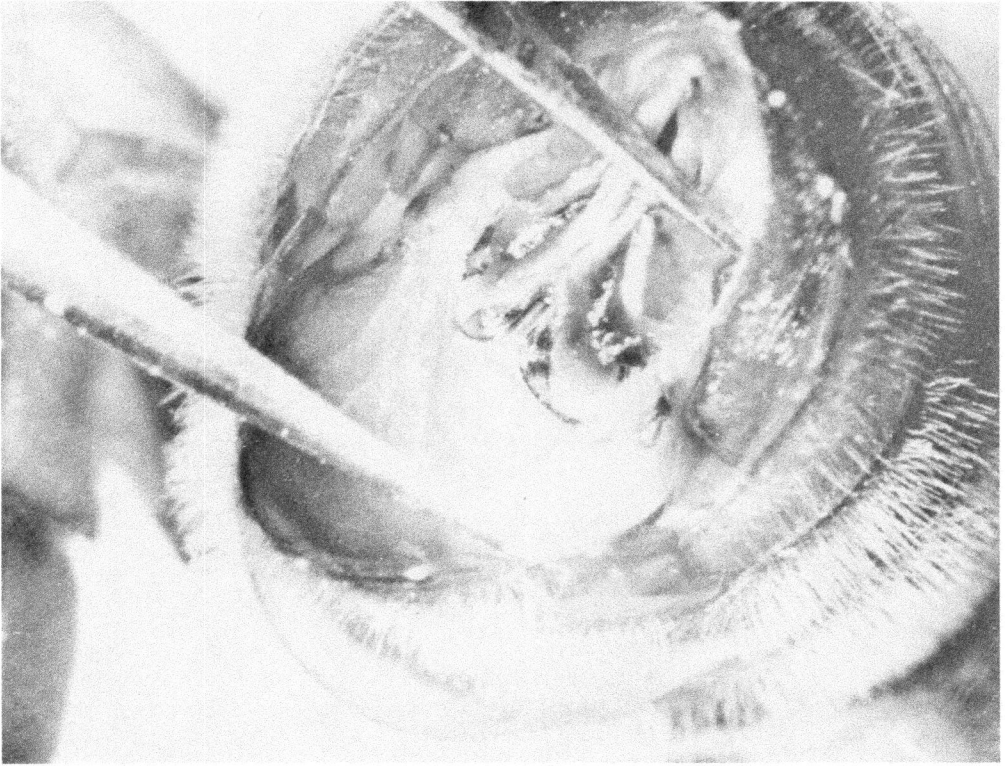

39. OPENING STING CHAMBER

The dorsal and ventral plates of the sting chamber are spread exposing the base of the sting.

40. POSITIONING STING HOOK

The dorsal, or sting, hook is brought over the queen and lowered into position against the base of the sting to fit between the bases of the sting lancets. Since the anaesthetization chamber and the opening hooks are attached to the same bar, no lateral adjustment of the opening hooks is necessary. The abdomen of the queen can be moved with the forceps to assist in fitting the hook against the sting base. Pressing the sting shaft down slightly with the dorsal forcep leg raises the sting base for easier positioning of the sting hook.

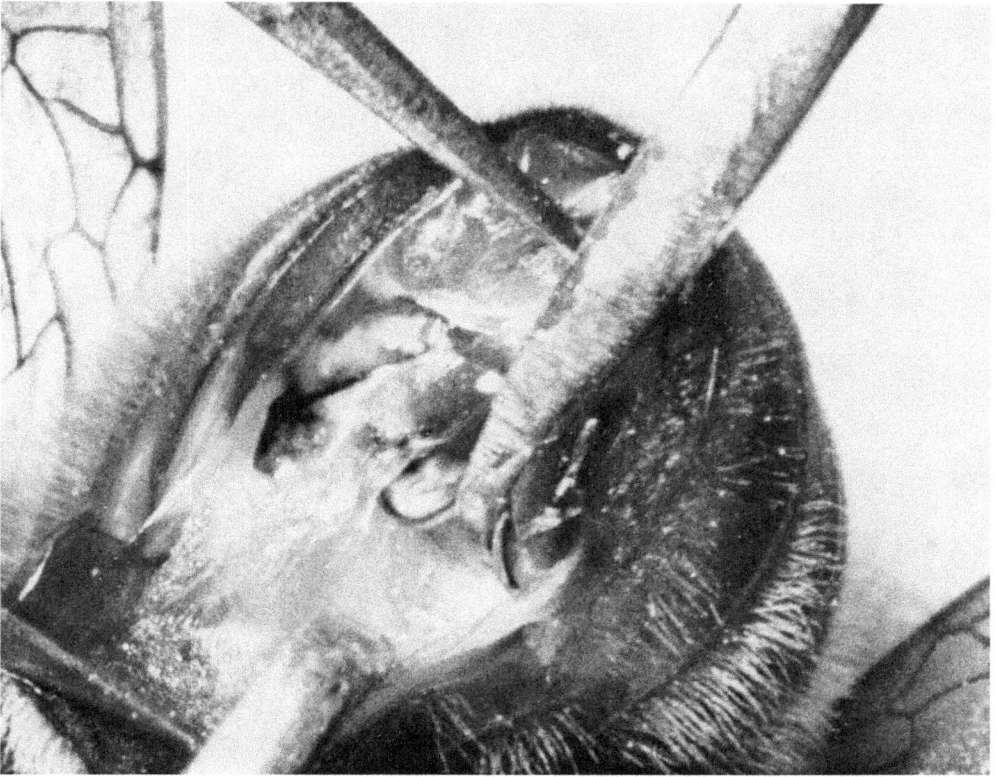

41. POSITIONING THE VENTRAL HOOK

The ventral hook is now brought over the queen and lowered into the sting chamber against the ventral plate until the tip touches the anterior membrane of the sting chamber.

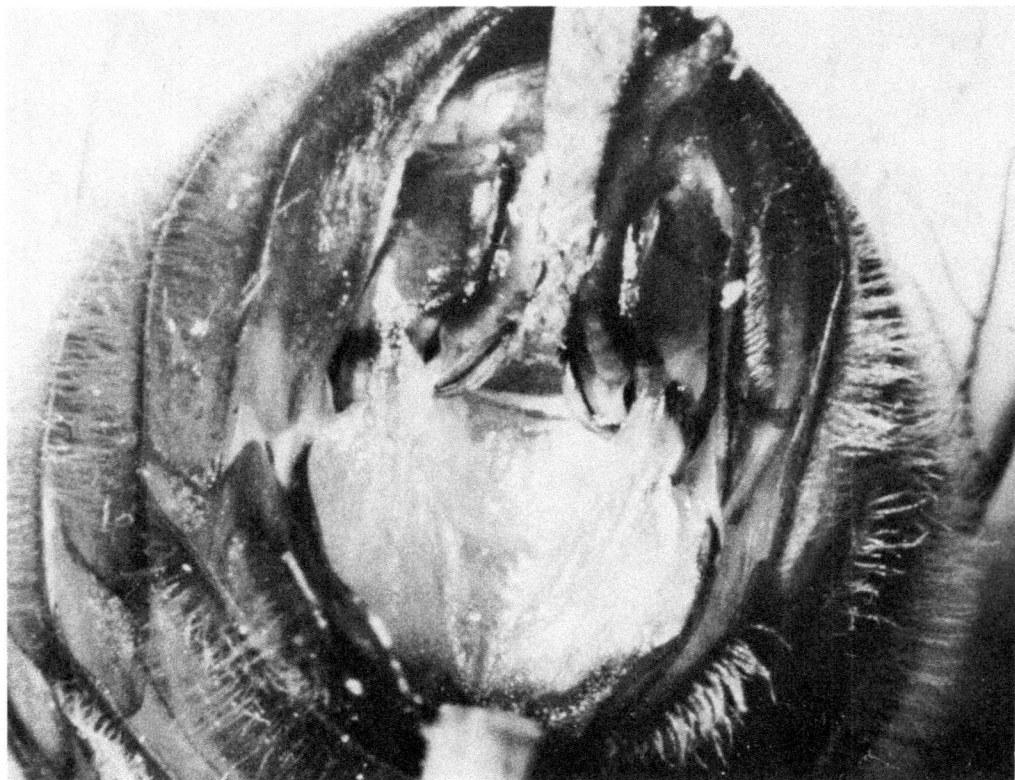

42. POSITIONING THE HOOKS

Hooks properly positioned.

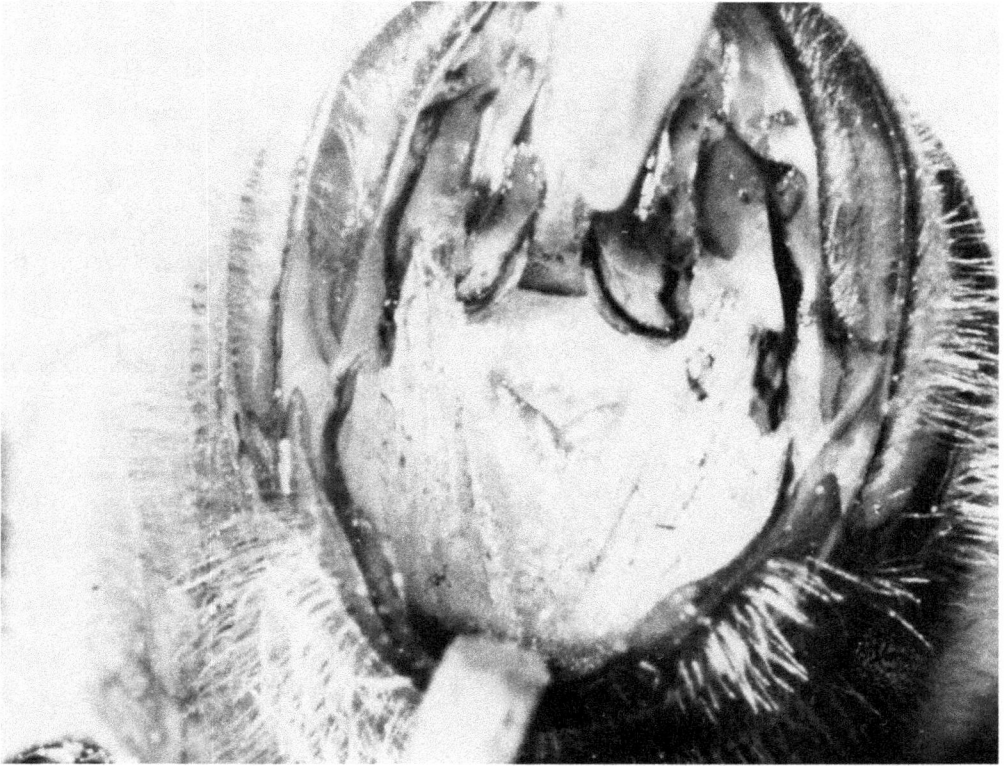

43. POSITIONING THE HOOKS

Stained vaginal orifice partly opened by adjustment of dorsal and ventral hooks.

Note: Flooding the sting chamber with plunger solution, or clear water, will cause the vaginal tissues to separate and will bring the vaginal orifice into relief. Use for study **only.**

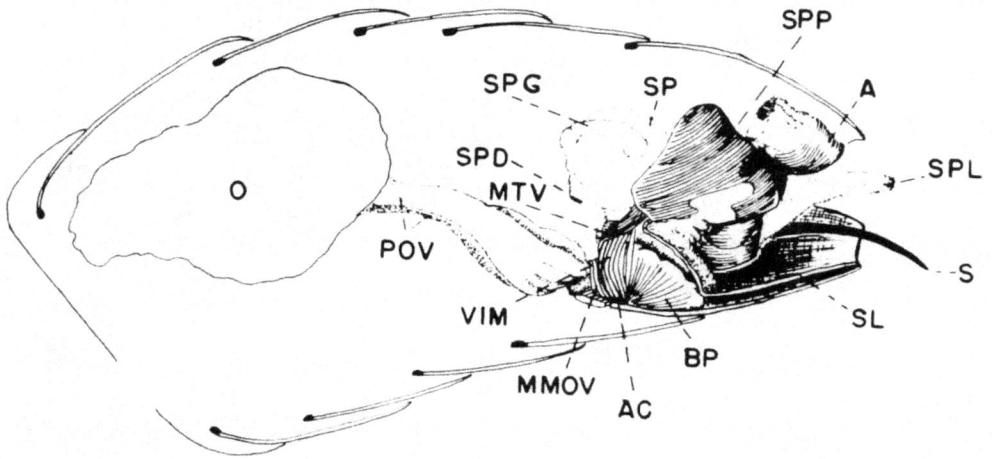

44. QUEEN STRUCTURE

The end of the syringe tip should be placed into the vagina and at least against the orifice of the median oviduct, and the semen injected into the lateral oviducts. Over a period of 15 to 24 hours the sperm move into the spermatheca as the queen slowly forces the semen out of the reproductive tract into the sting chamber. This lateral view of the queen's reproductive system shows the general relationship of structures involved in insemination. A anus, AC antecosta of sternum, BP bursal pouch, MMOV muscles of median oviduct, MTV tergo-ventral muscles, O ovary, POV paired or lateral oviducts, S sting, SL reflected lining of sting chamber, SP spermatheca, SPD spermathecal duct, SPG spermathecal gland, SPL sting palps, SPP spiracle bearing plate of sting, VIM ventral internal median muscle. (From Laidlaw, *J. Morph.*, 1944)

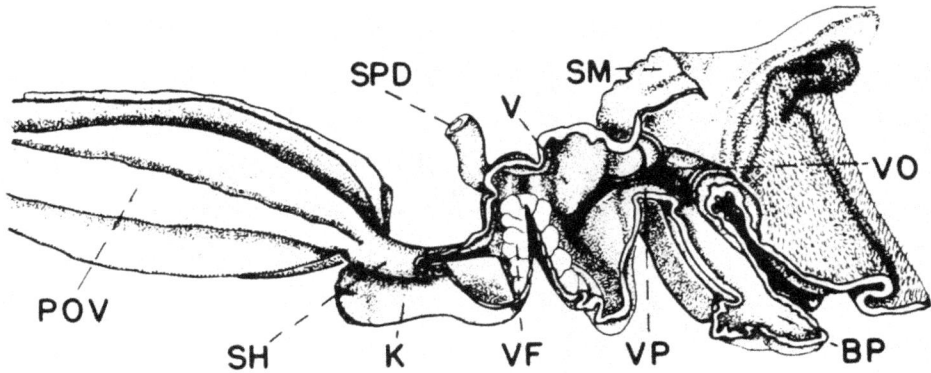

45. QUEEN STRUCTURE

To get semen into the lateral oviducts it is necessary to bypass the valve-fold, a structure arising as an invagination from the ventral side of the vagina at its junction with the median oviduct. This distended lateral view of reproductive tract illustrates the valvefold in relation to the vagina and the median oviduct. BP bursal pouch with side removed, K "Keel" of median oviduct, POV paired oviduct, SH "shelf" of median oviduct, SM sting membrane, SPD spermathecal duct, V vagina, VF valvefold, VO vaginal orifice, VP vaginal passage. (From Laidlaw, *J. Morph.,* 1944)

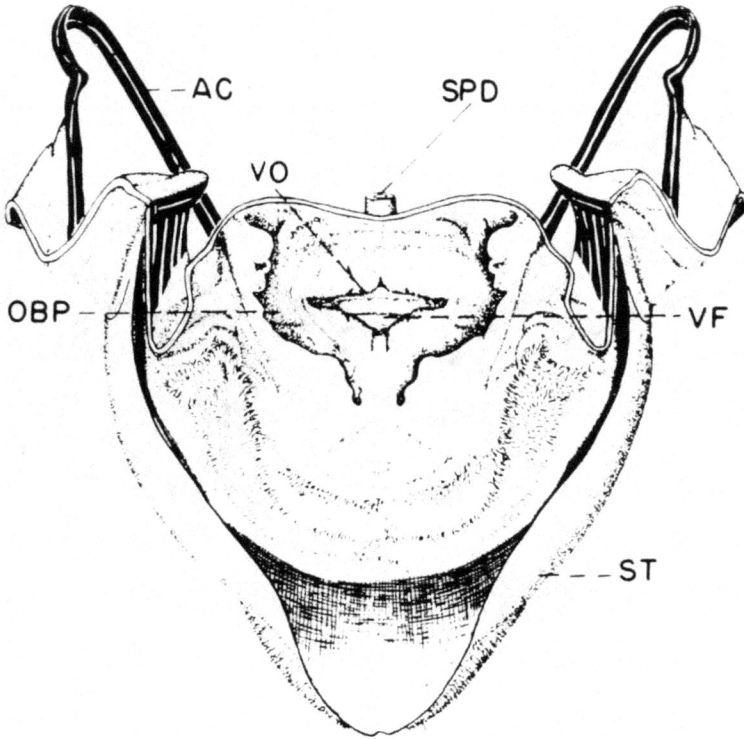

46. QUEEN STRUCTURE

In the normal position of the reproductive system, the valvefold occludes the opening of the median oviduct, and appears in the partly opened vagina as illustrated in this drawing. AC antecosta of sternum, OBP orifice of bursal pouch, SPD spermathecal duct, ST sternum, VF valvefold, VO vaginal orifice. (From Laidlaw, *J. Morph.*, 1944)

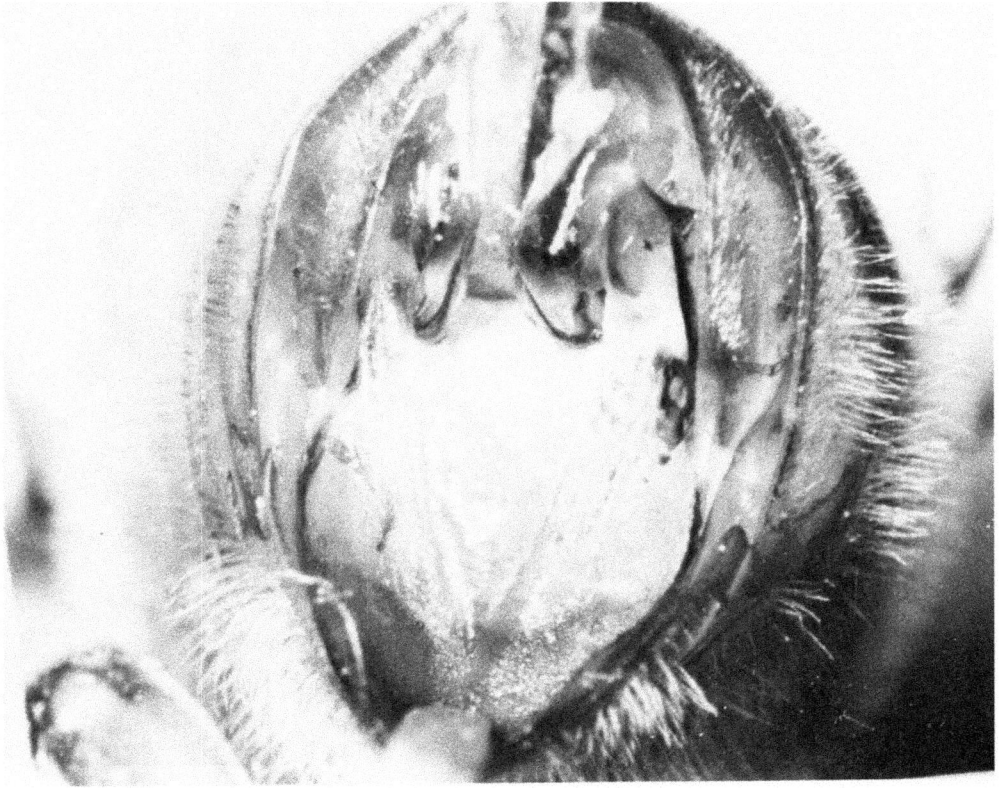

47. BYPASSING VALVEFOLD

The valvefold may be bypassed by the syringe in either of two ways: by pushing the valvefold ventrally with a flattened probe or with the end of the syringe tip, or by stretching the vaginal orifice so it is pulled partly or completely dorsal to the tip of the valvefold. The latter method is faster and less likely to result in injury to the queen by less skilled operators. This photograph shows the vaginal orifice pulled dorsally to almost beyond the valvefold, and also the "V" formed at the ventral side of the vaginal orifice.

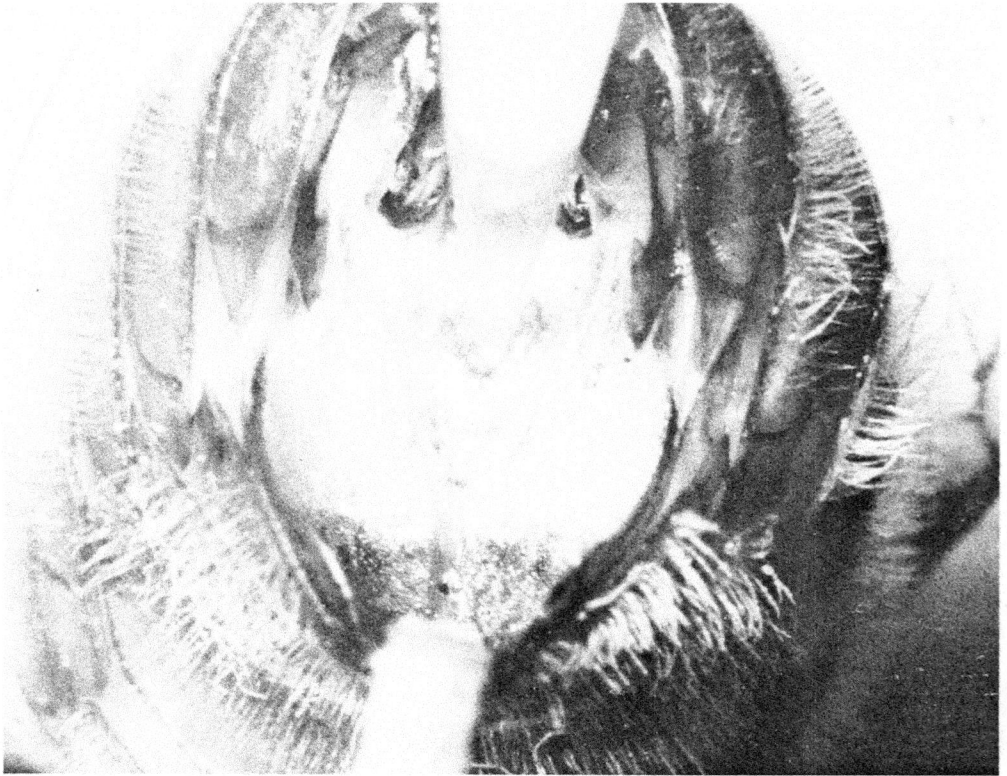

48. BYPASSING VALVEFOLD

The syringe is positioned directly above the opened vagina and dorsal to the end of the valvefold by moving the queen manipulator and queen over the microscope stage.

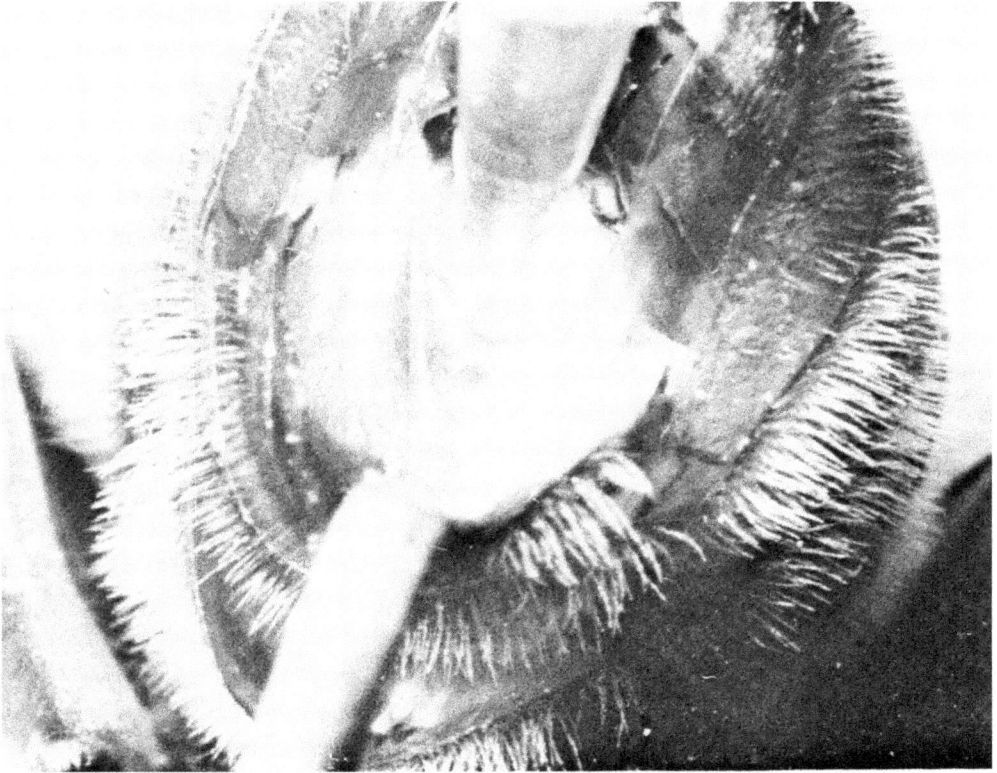

49. BYPASSING VALVEFOLD

The unstained vaginal orifice may sometimes appear as a small cavity in the anterior sting chamber membrane. The tip of the syringe should be inserted directly into this cavity, after a slight amount of the fluid taken into the syringe after the semen, is ejected as a lubricant. Often the orifice is only slightly visible, if at all, in which case the syringe is inserted just dorsal to the bottom of the "V" formed by the stretched membranous fold that normally partially overlies the vaginal orifice ventrally.

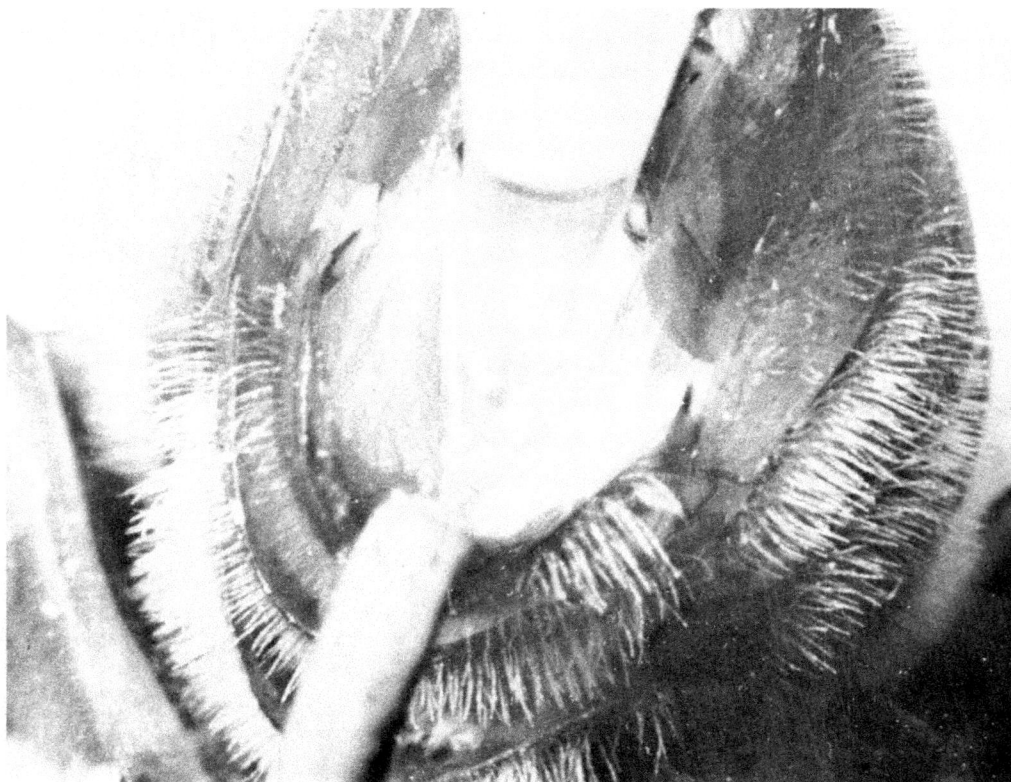

50. INSERTING SYRINGE

The syringe is inserted **no farther than necessary,** possibly about 1 mm or until the tissue around the syringe moves with the syringe. The syringe should **slip** easily into the vagina without movement of the surrounding tissue.

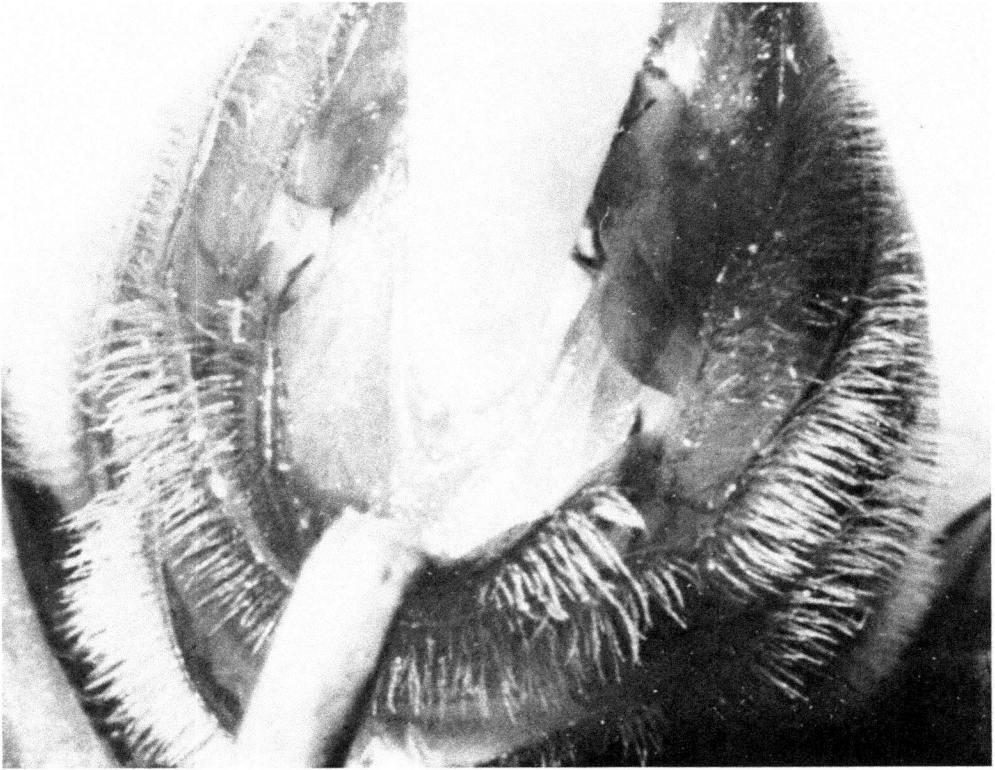

51. INJECTING SEMEN

First turn the plunger **slightly** clockwise to test whether the syringe is properly placed. If semen moves easily down the tip, injection can be rapid. If the semen merely starts but does not flow, either the syringe is improperly placed, a seal has formed over the end of the tip, or mucus has been taken into the syringe. Should this difficulty arise, the syringe should be withdrawn and movement of the semen in the syringe tested. If there is no seal or mucus, the syringe should be repositioned and inserted. Should slight leakage appear around the tip, insert it slightly deeper, but **never** retake into the syringe any semen that has leaked, because semen that has touched the sting chamber membranes may be contaminated. Leaked semen may be removed by swabbing with a bit of facial tissue held with forceps.

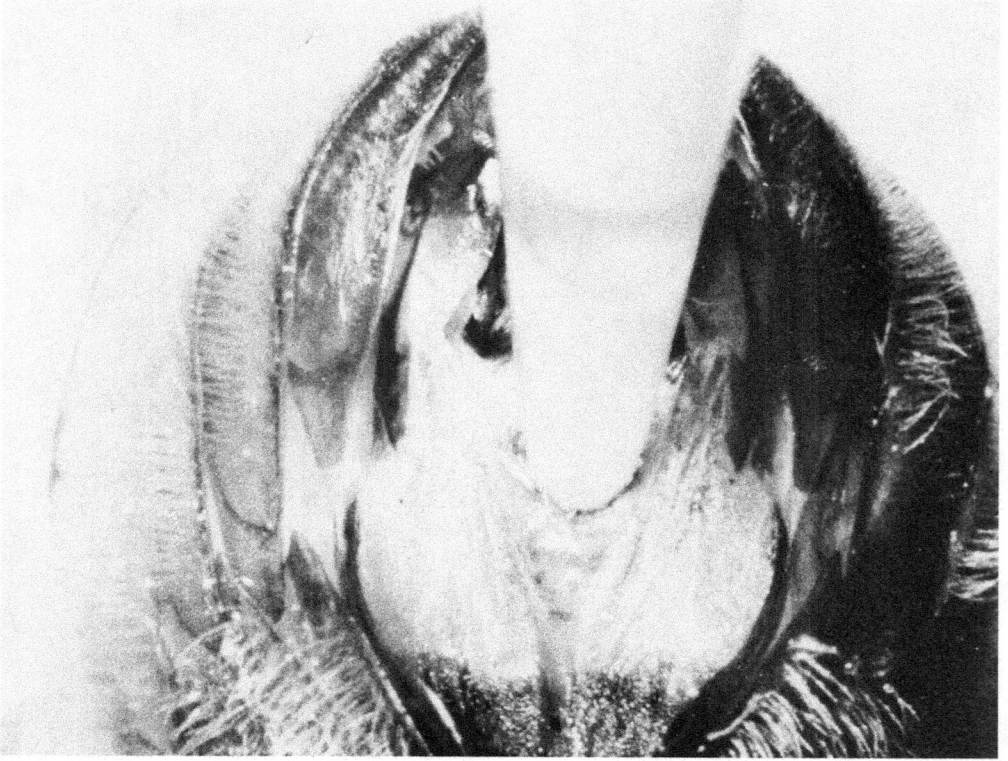

52. INJECTING SEMEN

Injection nearly complete. Inject all of the semen and all, or nearly all, of the air bubble; then raise the syringe to maximum height. Sometimes a droplet of semen may well up out of the vagina, but this is of no consequence.

When the syringe is raised, move the liquid plunger to the end of the tip to avoid the formation of a seal across the end of the tip.

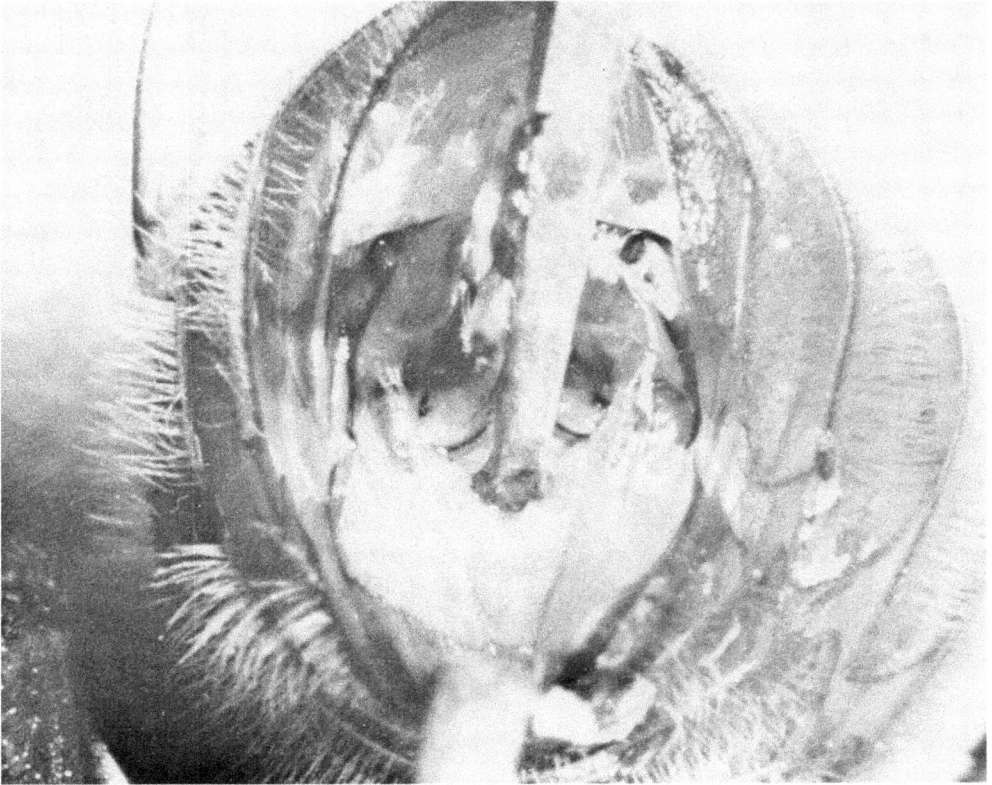

53. DISENGAGING HOOKS

The sting hook is disengaged by moving it inward and then up and back as far as it will go. The ventral hook is then removed in like manner.

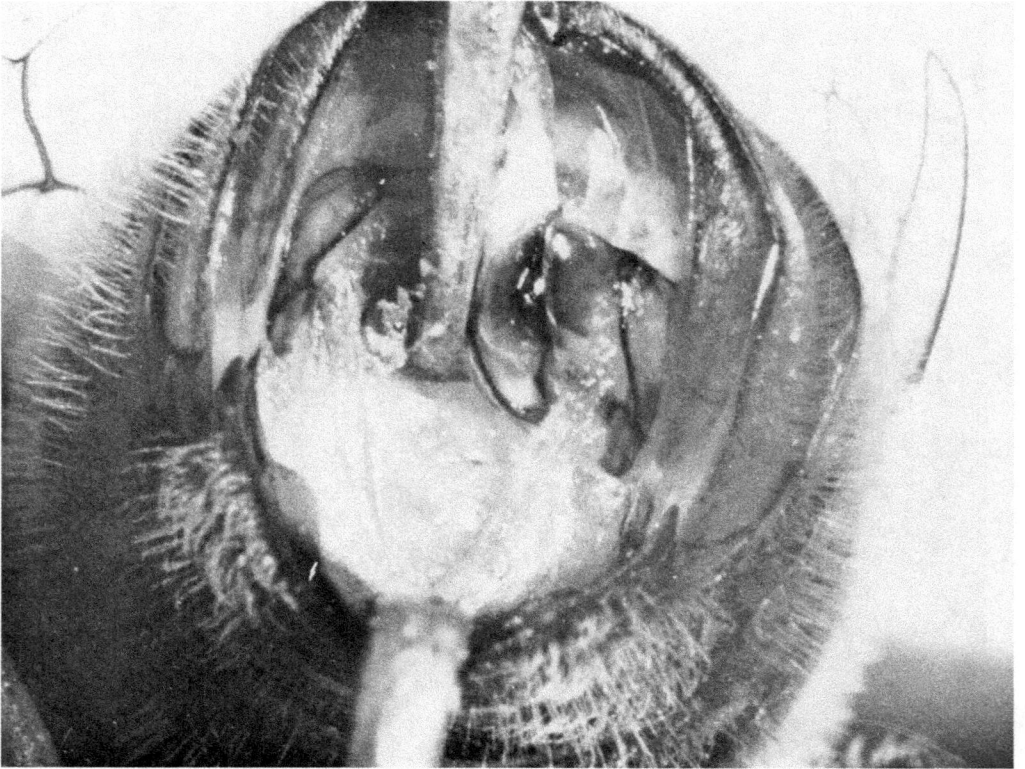

54. ALTERNATE WAY TO INSERT SYRINGE

An alternate way to insert the syringe is to lower the valvefold with a probe. When this is done, the sting chamber membranes are not stretched but the sting is moved dorsally only until the vaginal orifice opens part way.

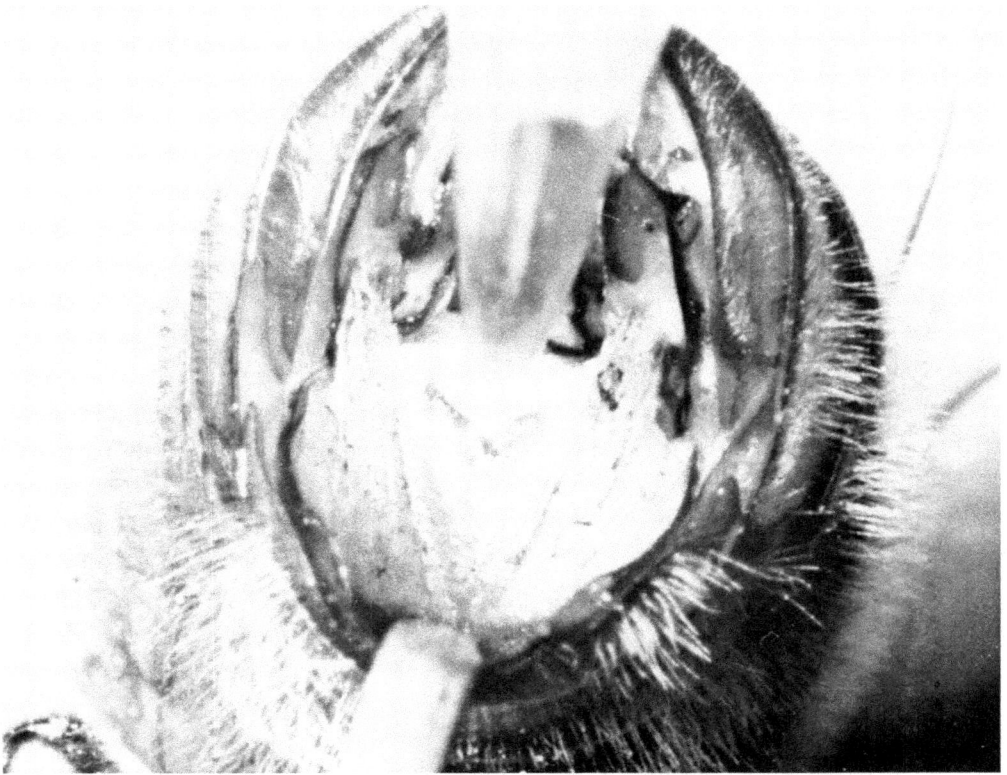

55. ALTERNATE WAY TO INSERT SYRINGE

The syringe is then positioned just above the partly opened vaginal orifice or just ventral to it because the vagina will be pulled ventrally with the probe. (Vagina stained)

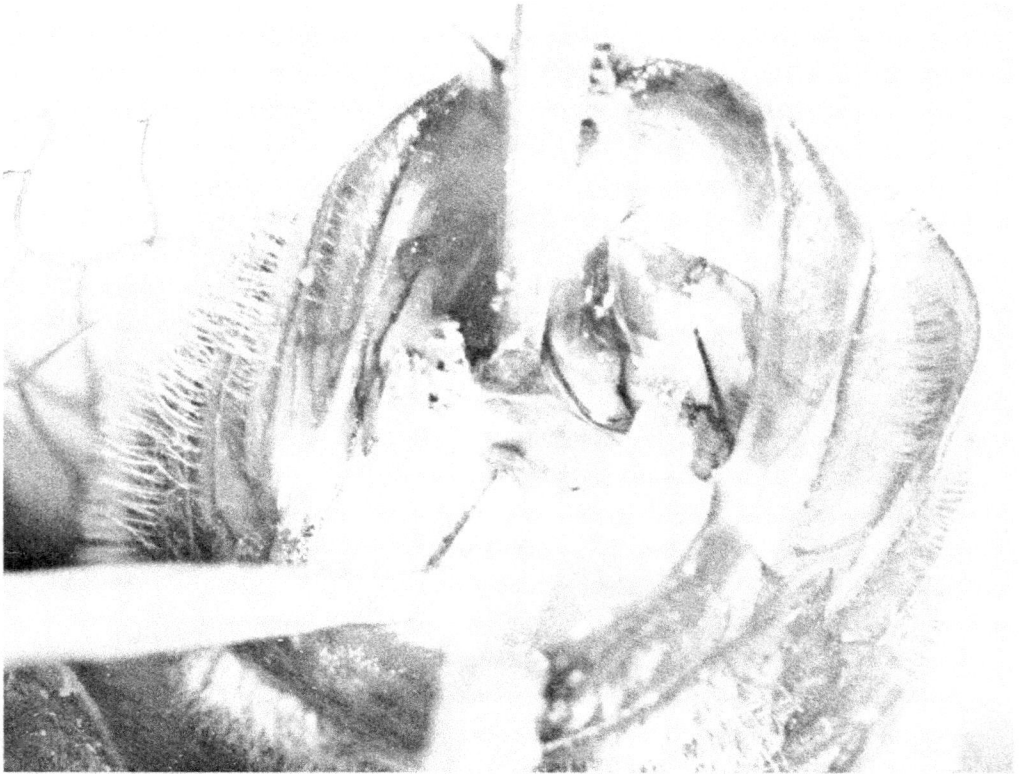

56. ALTERNATE WAY TO INSERT SYRINGE

The probe is steadied with tips of the fingers of the left hand lightly touching the side and sloping top of the anaesthetization chamber, and the end of the probe is placed near the sting hook. The probe is inclined about 5° toward the sternum.

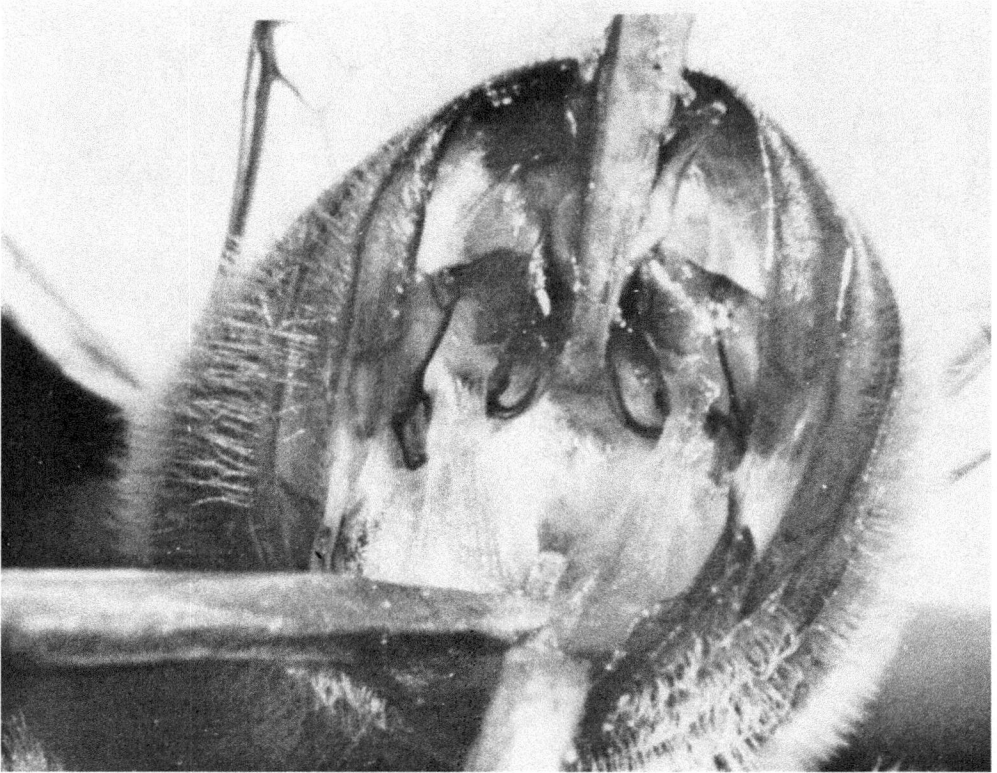

57. ALTERNATE WAY TO INSERT SYRINGE

It is then pressed into the sting chamber membrane and, maintaining the inclination, is moved ventrally which opens the vagina and pushes the valvefold from over the median oviduct orifice.

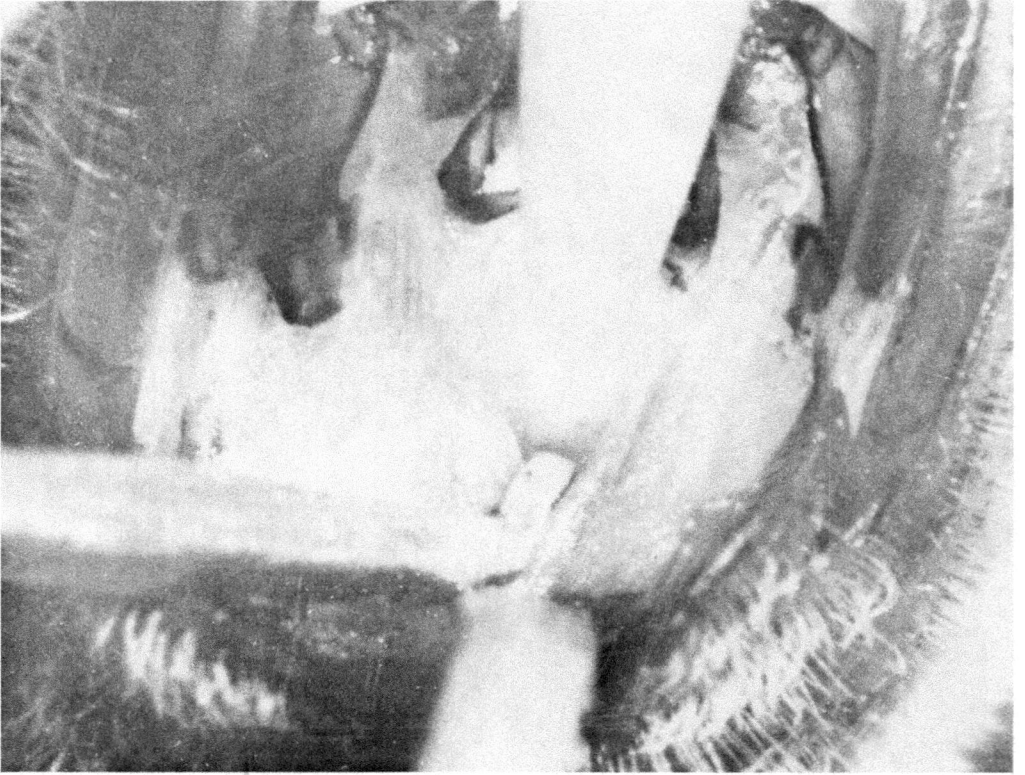

58. ALTERNATE WAY TO INSERT SYRINGE

The syringe is then inserted into the open vagina, which often appears as a deep cavity, and the probe is removed.

59. REMOVING QUEEN FROM
ANAESTHETIZATION CHAMBER

When the hooks have been disengaged and moved up and away from the queen, the queen manipulator is set off to the left side of the micro-scope and the queen removed by grasping a leg and opening the chamber.

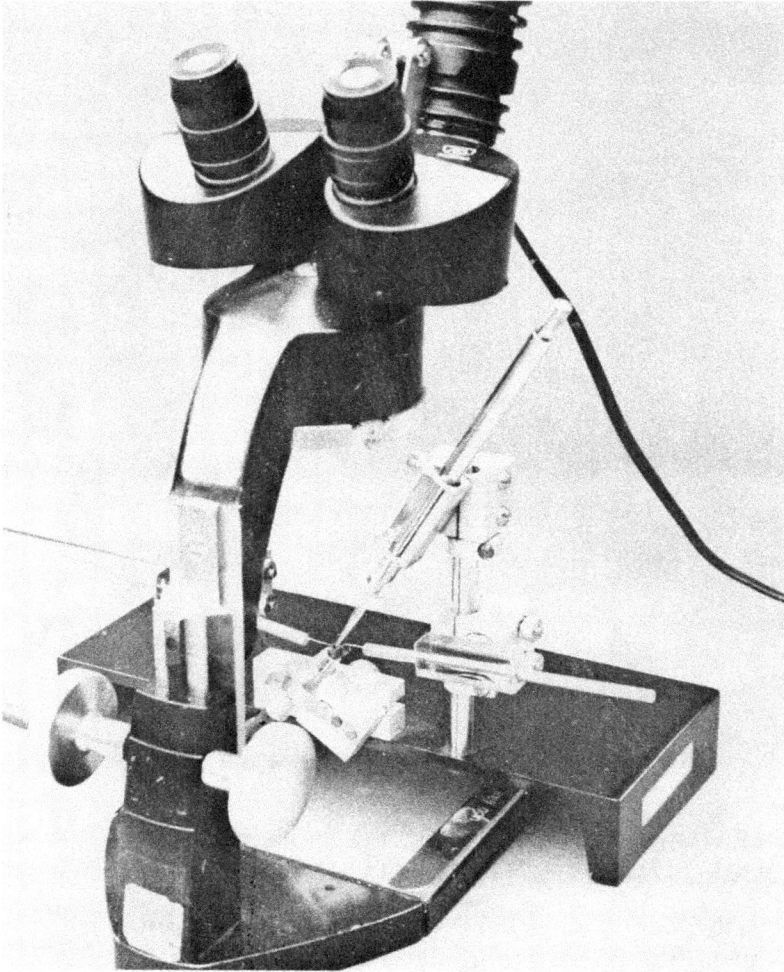

B. MACKENSEN INSTRUMENT

Preparation of the syringe, obtaining semen, and many other operations are similar for both the Laidlaw and Mackensen instruments and the photographs pertaining to these are not repeated.

60. THE MACKENSEN INSTRUMENT AND MICROSCOPE

The instrument is shown across the stage of a stereoscopic dissecting microscope which is equipped with a 1X objective and 12X or 15X occulars. Lamp has a heat filter. Queen is held in a tube to which CO_2 is conducted through flexible tubing. The sting chamber is opened and the sting pulled dorsally by hooks secured to the instrument by rectangular boxes that are movable up and down. The boxes are provided with leaf springs and are attached to bases that swivel on posts. The syringe is held by a similar mechanism.

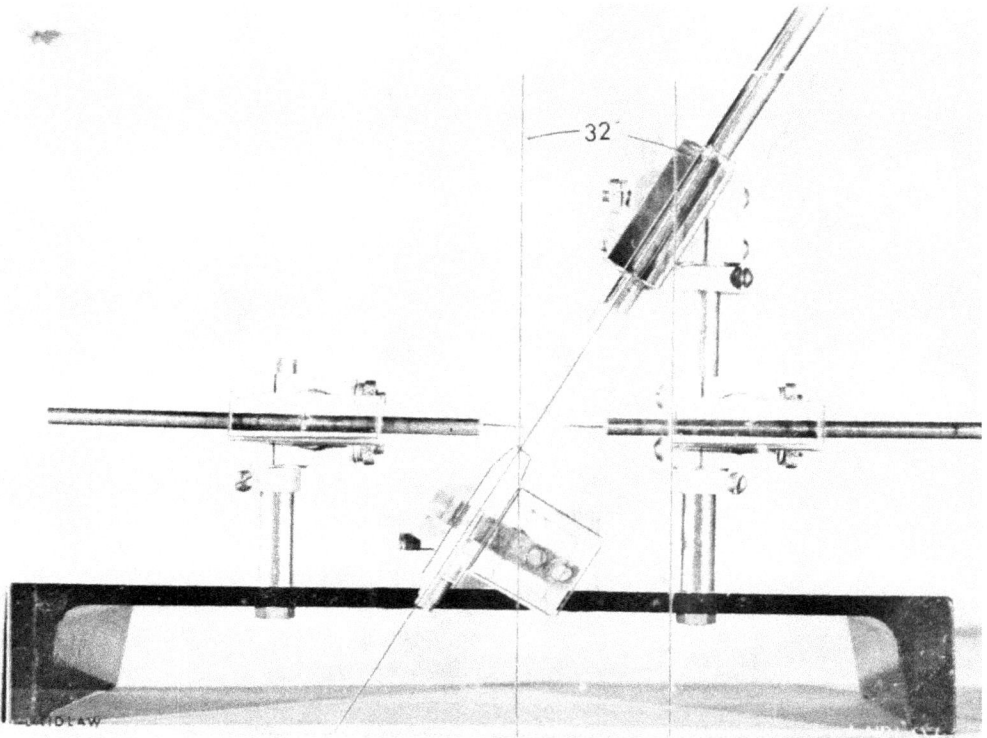

61. MACKENSEN INSTRUMENT AND ADJUSTMENT GUIDE

Proper alignment of the Mackensen instrument is necessary for fast and easy inseminations. There is some tolerance in the adjustment of both the Mackensen and Laidlaw instruments, but the unskilled would do well to adhere to the alignments recommended.

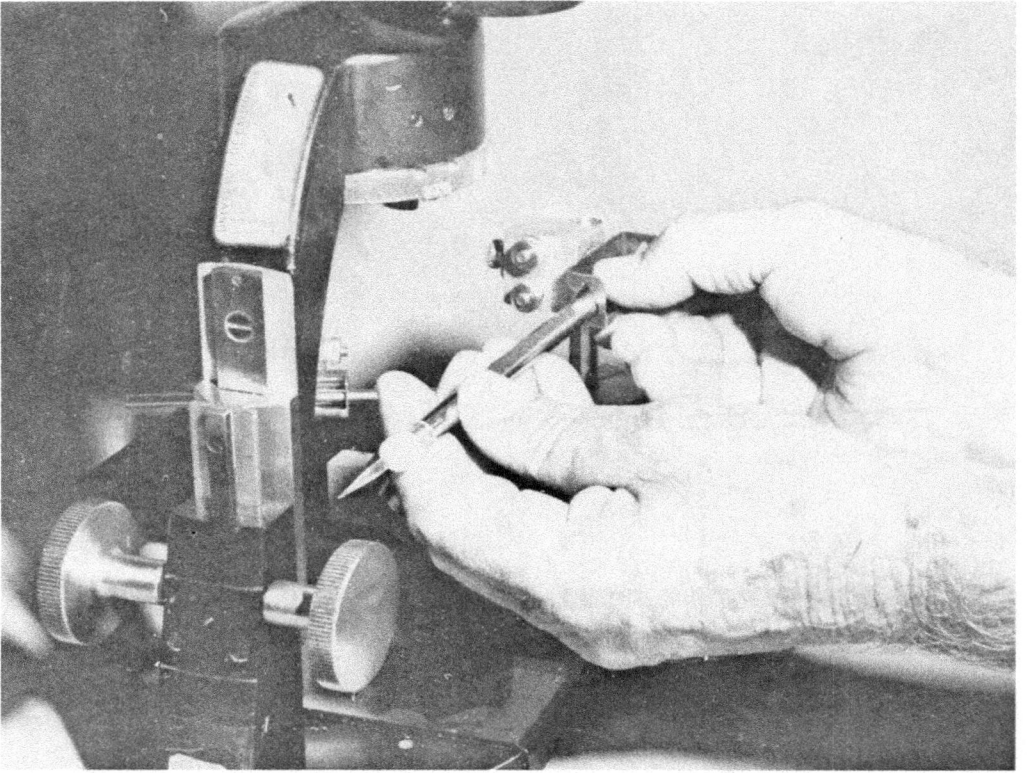

62. FASTENING SYRINGE IN SYRINGE HOLDER

The screw plunger end of the prepared syringe should be slipped into the syringe holder and pushed through to the adapter.

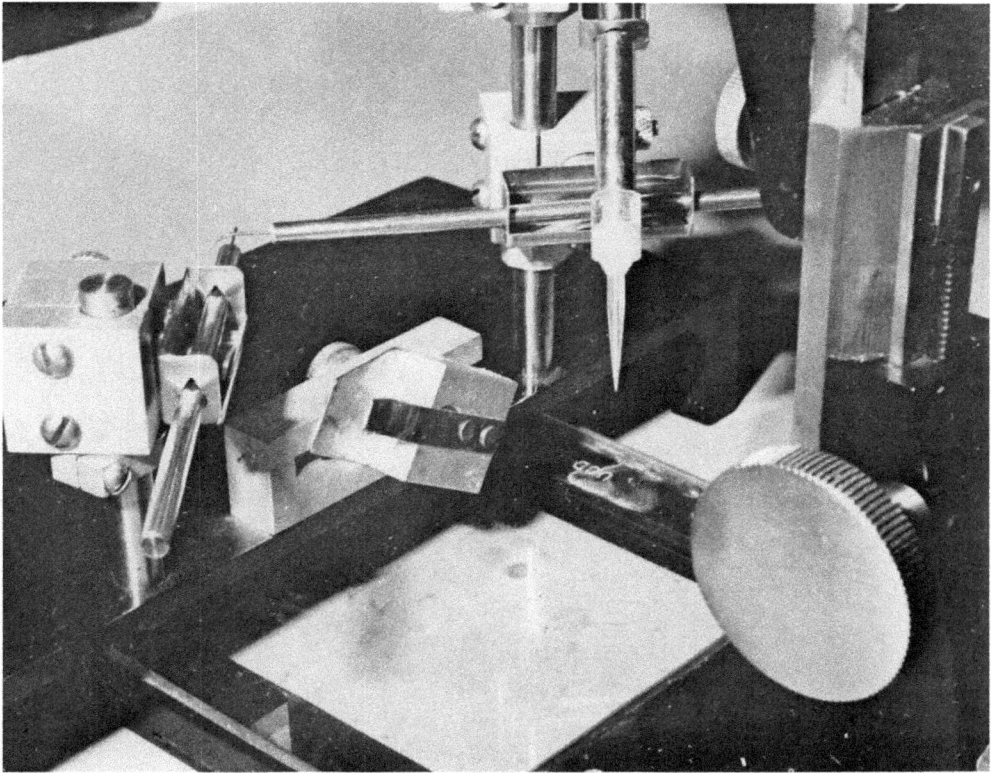

63. POSITIONING SYRINGE FOR SEMEN UPTAKE

The instrument is moved away from the operator about 5 centimeters to give room to bring the everted drone to the syringe tip and observe the taking of the semen.

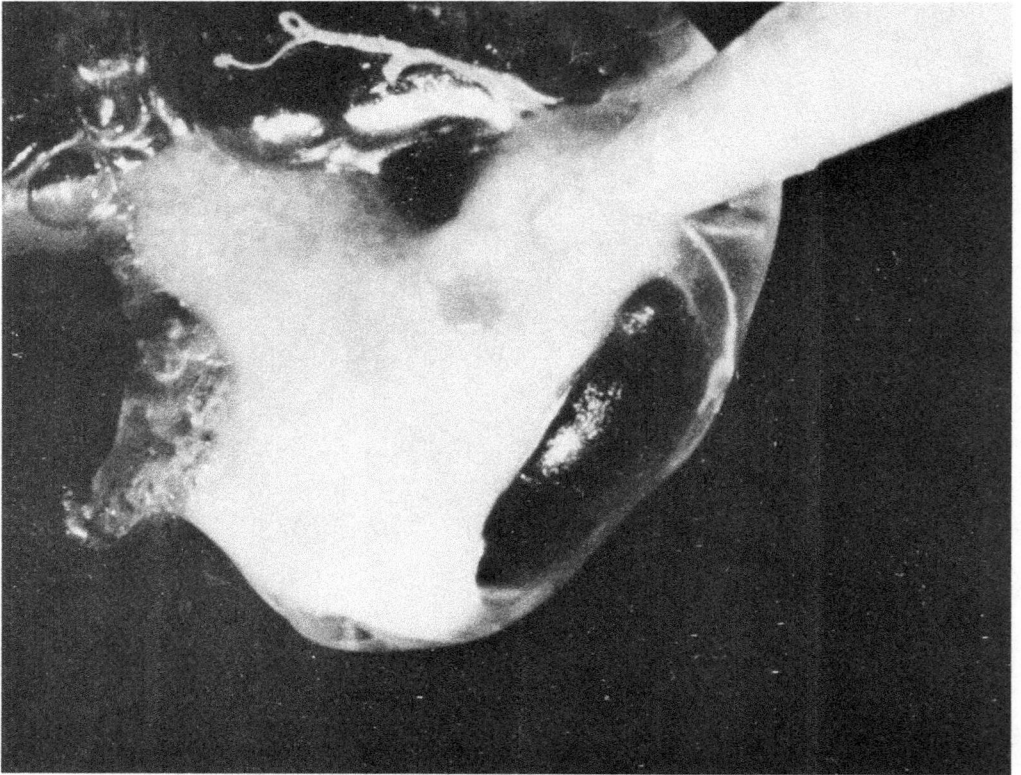

64. LOADING SYRINGE

The ejaculated drone is brought into focus under the microscope and the syringe tip is slipped down **to just touch the semen.** The process of loading the syringe is the same as with the Laidlaw apparatus.

65. LOADING SYRINGE

If contact of the tip with semen is lost, a small amount of semen should be ejected and this touched to the semen on the penis to regain contact. The same procedure should be used in beginning to take semen from the next drone.

Between each drone, withdraw the semen a millimeter or more into the syringe to prevent the formation of a seal at the end of the tip.

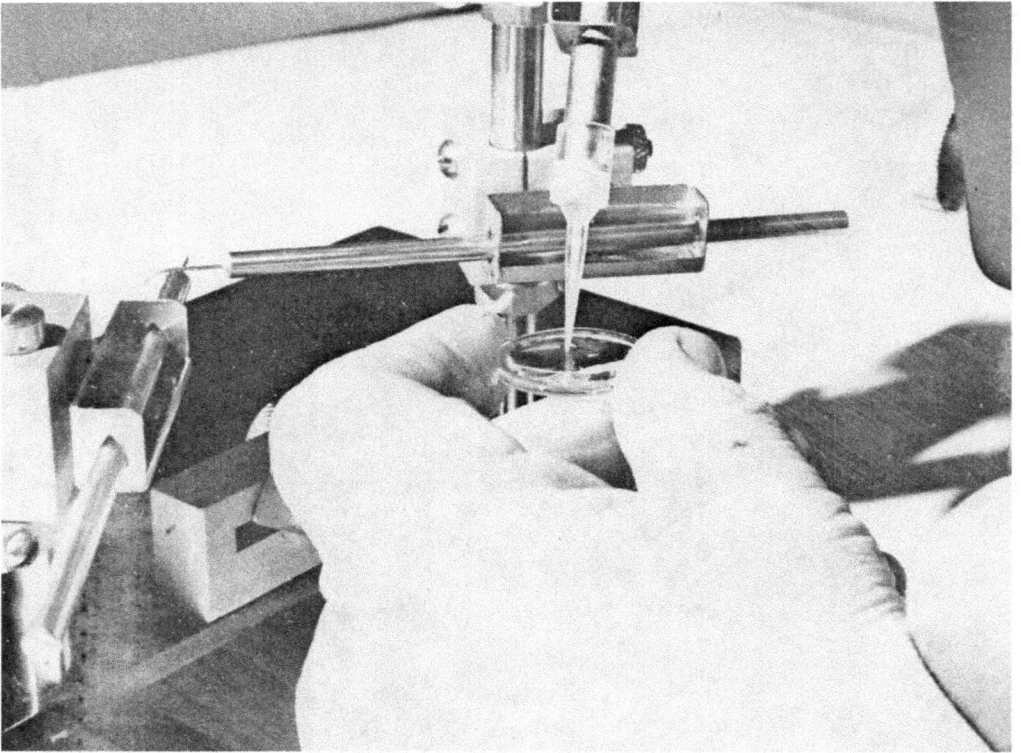

66. ANTIBIOTIC

About ½ cubic millimeter of the plunger fluid is taken into the syringe after it is loaded with semen.

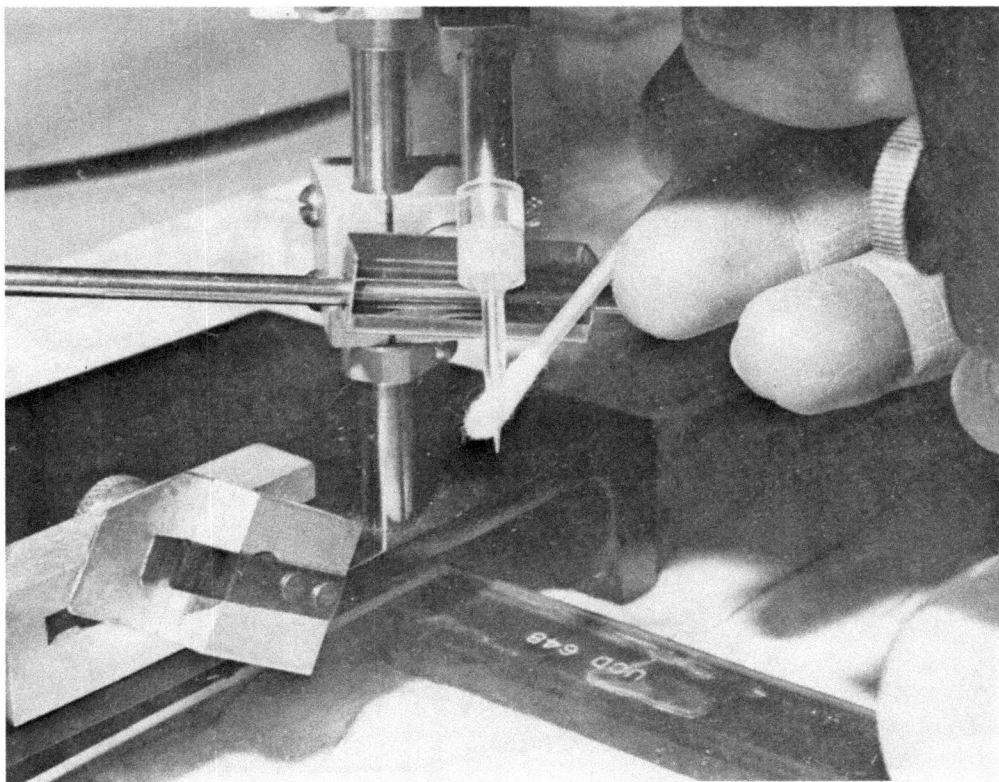

67. WIPING SYRINGE TIP

Following loading, the tip is cleaned externally with clean facial or laboratory tissue or a cotton swab dipped in the plunger fluid. Discard swab or tissue after one use.

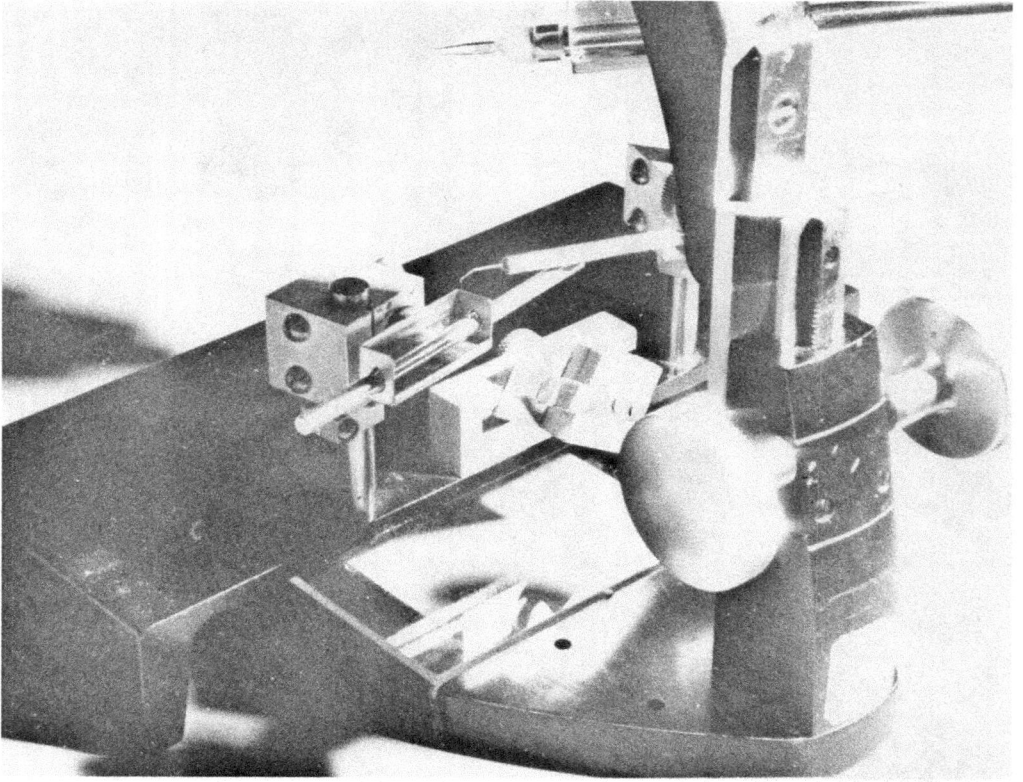

68. PREPARING TO PLACE QUEEN IN INSTRUMENT

The hooks and loaded syringe tip are moved away from the operator so as not to interfere with subsequent operations, and the instrument is slipped toward the operator to bring the queen holder block into the microscope field.

69. ADJUSTING CO$_2$ FLOW

The CO$_2$ flow is adjusted until adjacent bubbles emerge from the queen stopper when it is submerged in a beaker of water.

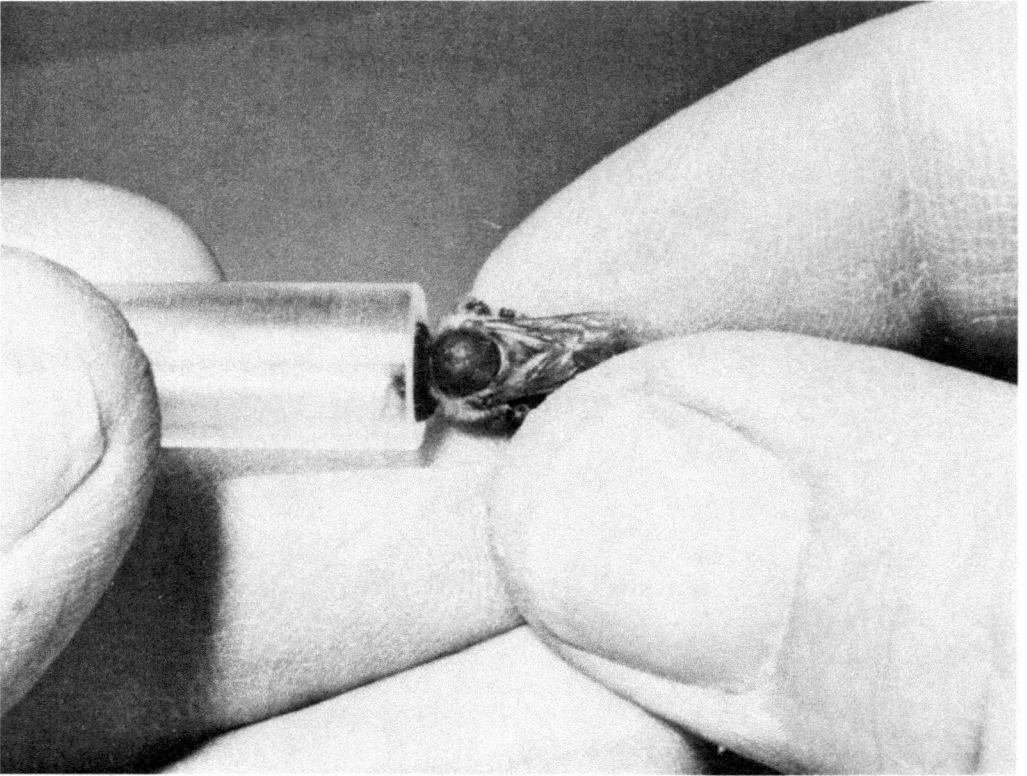

70. PUTTING QUEEN IN HOLDING TUBE

The anaesthetized queen is taken from the cage in the anaesthetization container and is pushed head first into the back-up tube which is supplied with the Mackensen instrument.

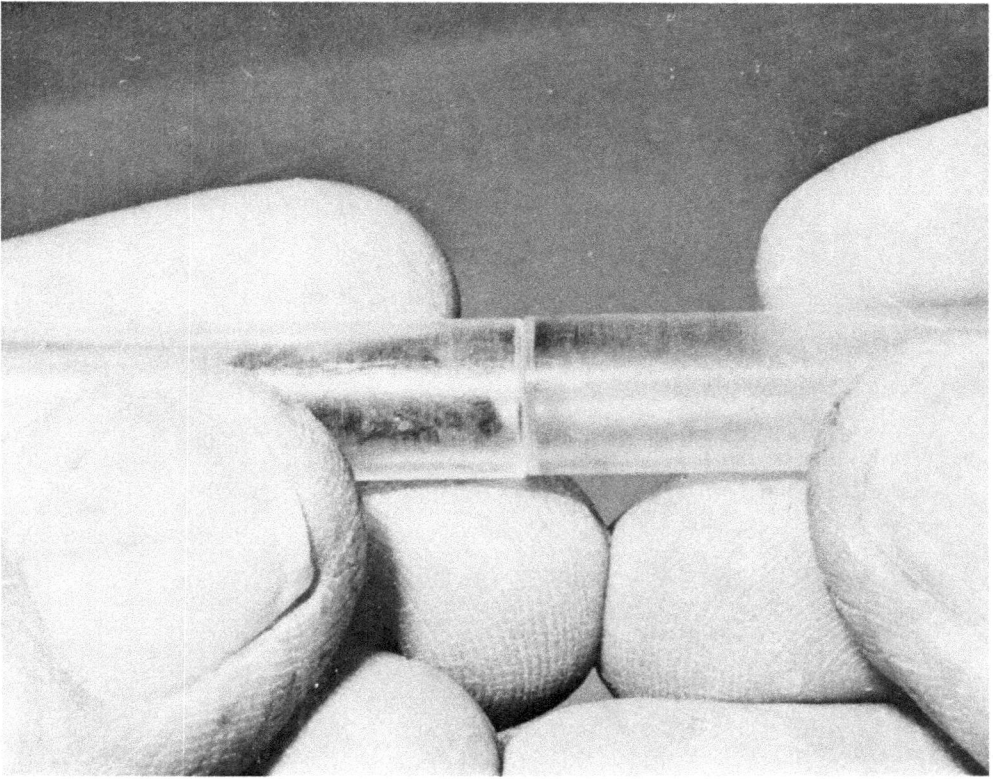

71. PUTTING QUEEN IN HOLDING TUBE

The large opening of the holding tube is butted against the back-up tube and the queen is blown from the back-up tube into the holding tube.

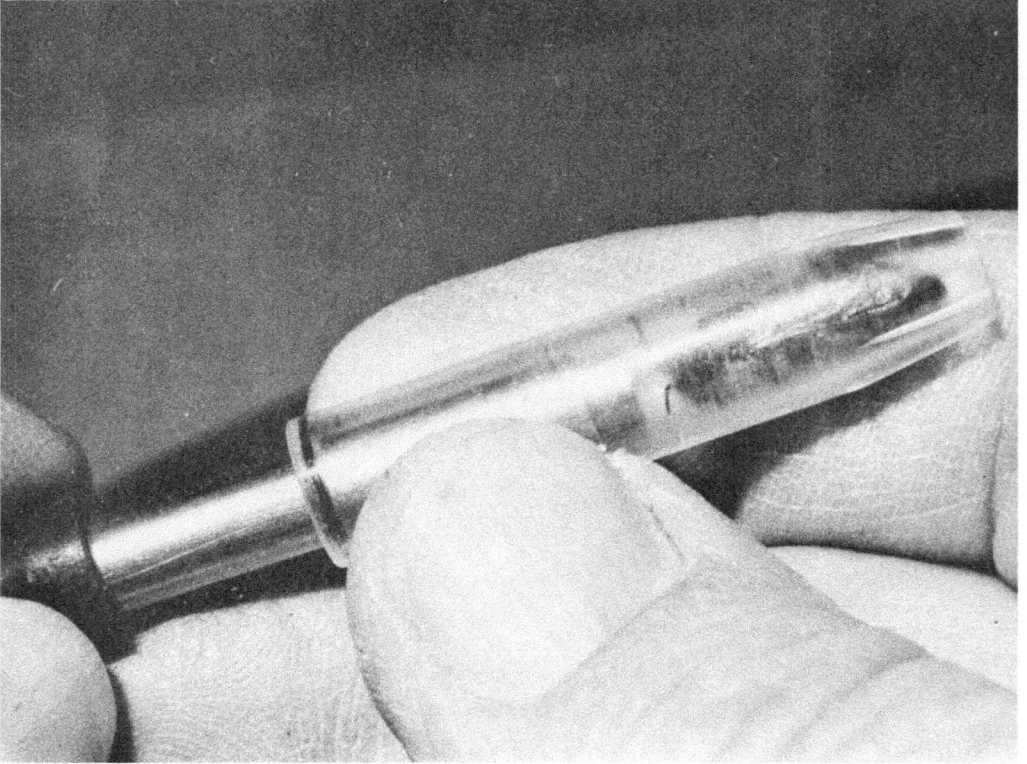

72. PUTTING QUEEN IN HOLDING TUBE

The tube stopper through which the carbon dioxide flows onto the queen in the tube is inserted into the holding tube.

Note: Regurgitation of food by the anaesthetized queens causes the interior of the holding tube to become sticky. The tube should be cleaned frequently with a cotton swab dipped in water.

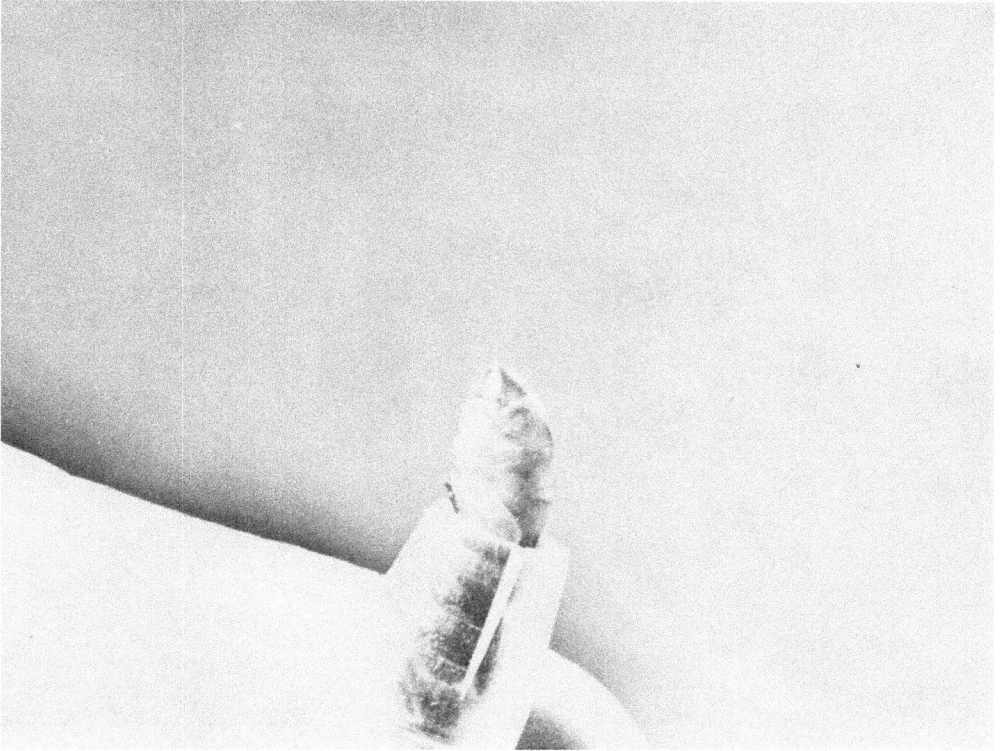

73. PUTTING QUEEN IN HOLDING TUBE

The queen is pushed by the stopper until 4 or 5 abdominal segments protrude beyond the tube end.

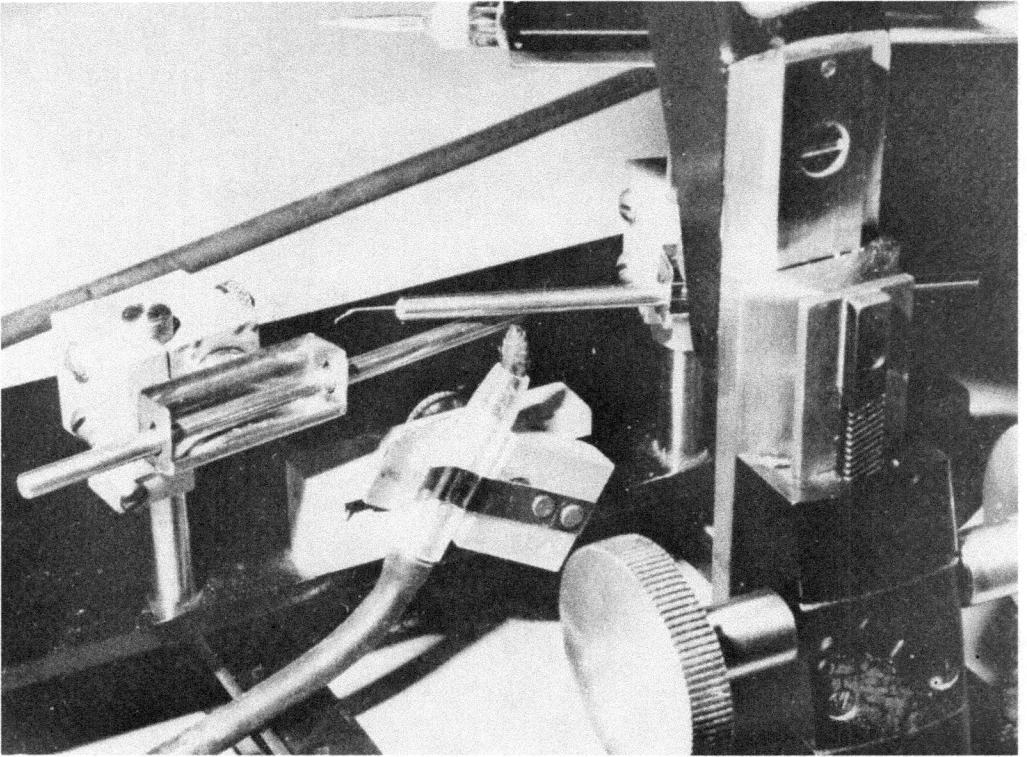

74. PLACING THE HOLDING TUBE
IN THE INSTRUMENT CLIP

The holding tube with queen is fastened in the clip so the base of the taper of the tube is aligned with the top of the holding block and the ventral side of the queen is to the left of the operator.

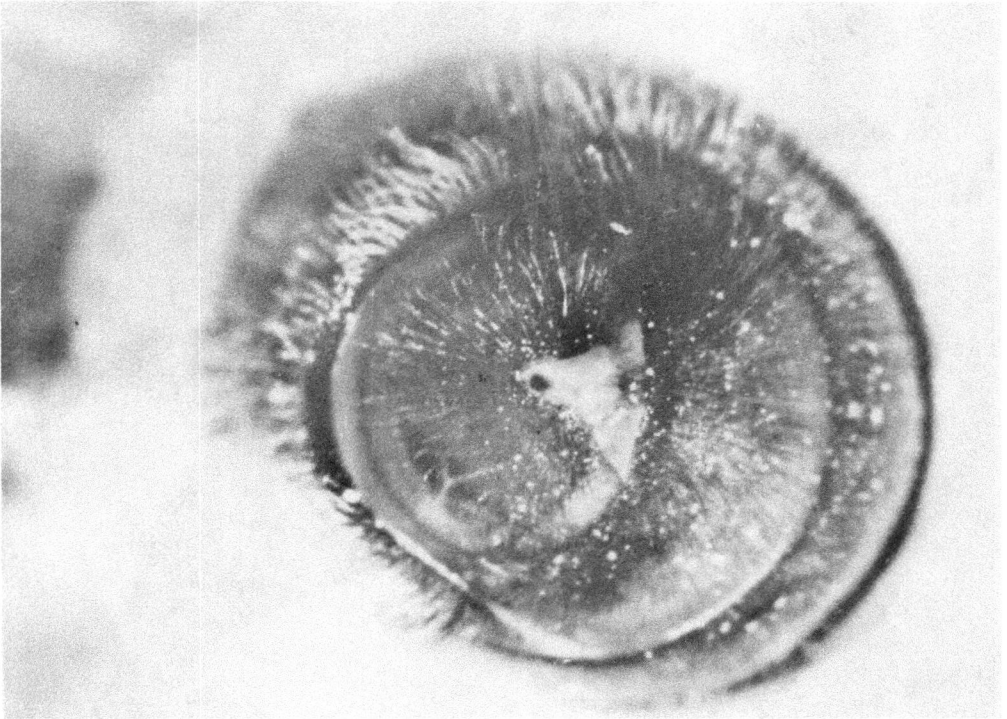

75. POSITIONING QUEEN IN MICROSCOPE FIELD

The instrument is moved until the queen is in the center of the microscope field.

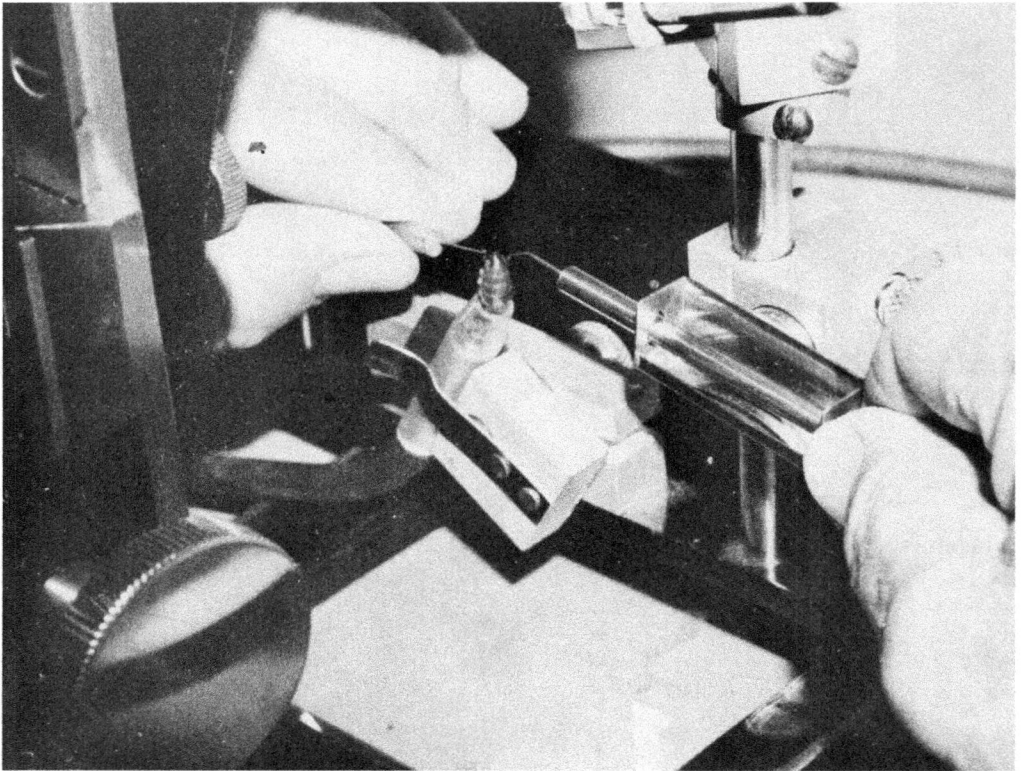

76. METHOD OF HOLDING OPENING HOOKS

The opening hooks slip toward and away from the queen on springs in rectangular boxes that are movable up and down and are attached to bases that swivel on posts. If properly adjusted, the movements of all parts are smooth. Greatest control over the in and out movement of the hooks, and also their lateral adjustment, is attained by holding the ventral hook handle near the hook itself, and the sting hook handle with the thumb and finger against the outer surface of the hook-holding box. By rolling the thumb and forefinger against each other, very precise movement of the sting hook is attained.

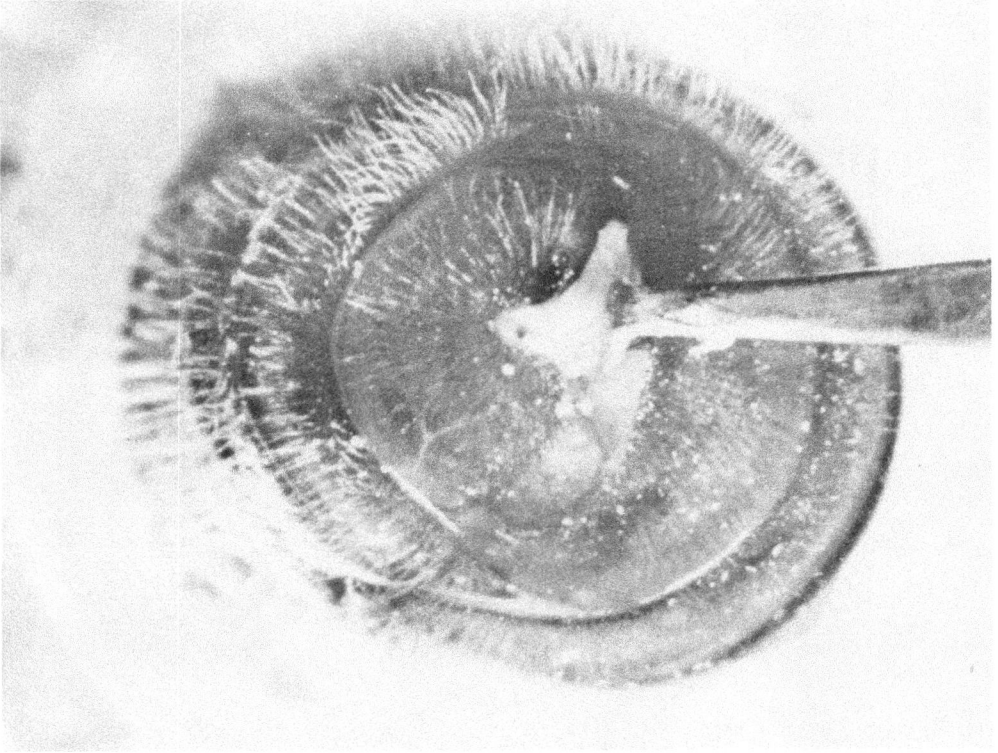

77. INSERTING STING HOOK

The sting hook is moved toward the queen and is inserted into the sting chamber. This is easy if the sting chamber is partly open. If the sting chamber is closed, the hook must be worked between the tergum and sternum into the chamber.

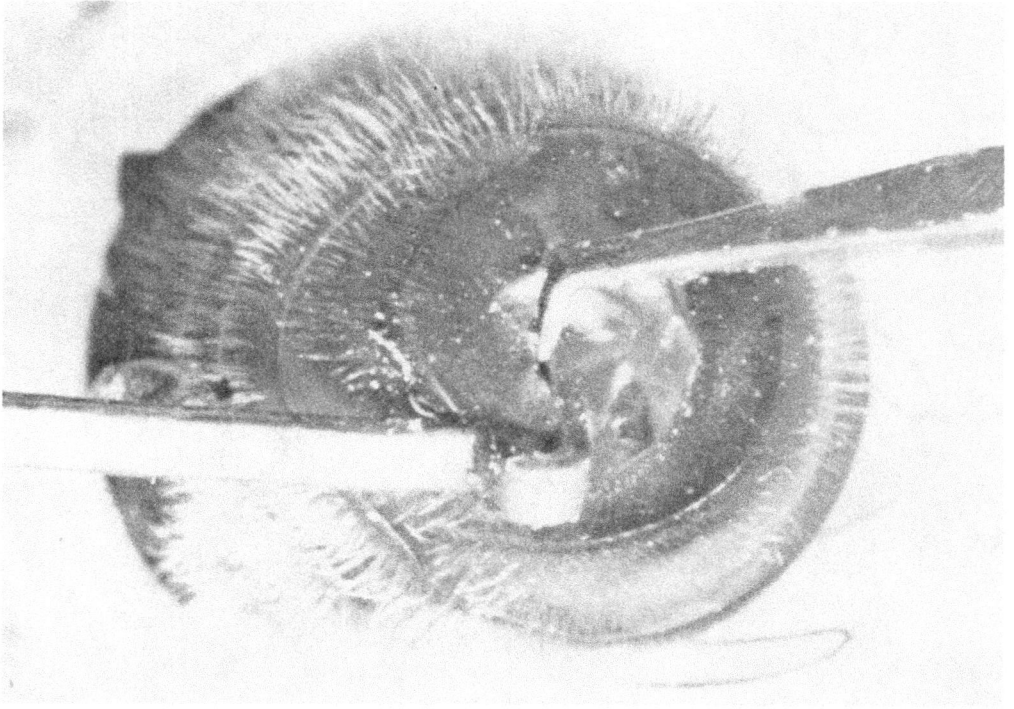

78. INSERTING VENTRAL HOOK

The sting hook is turned partly sideways and the ventral hook is brought toward the queen and is inserted into the sting chamber.

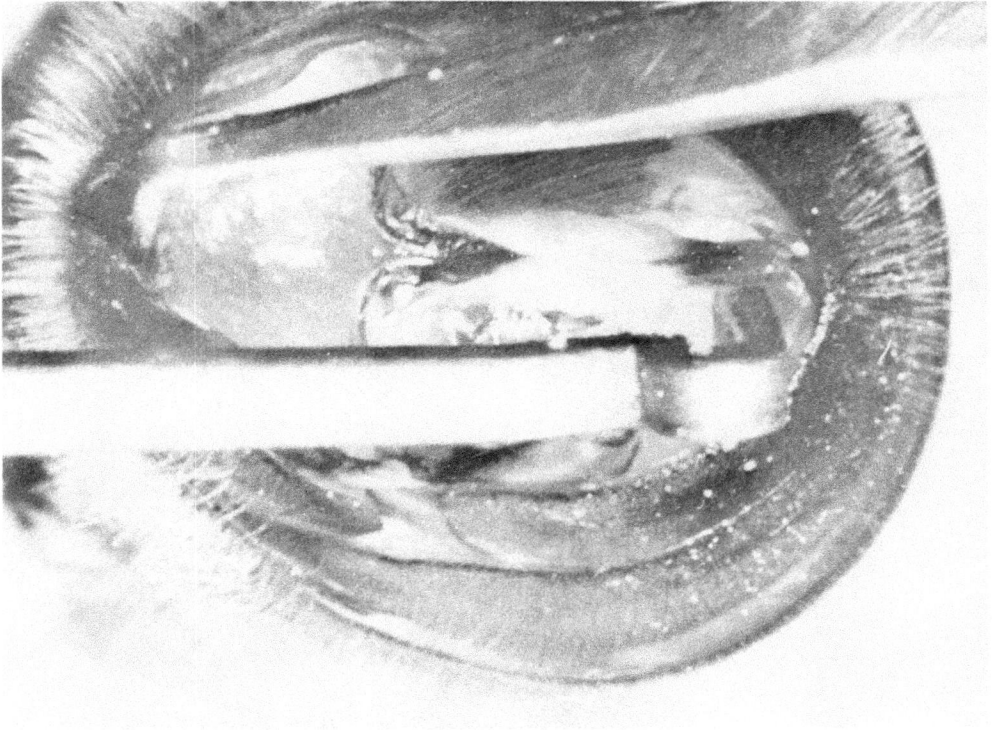

79. POSITIONING STING HOOK

The hooks are now pushed past each other and the sting hook is positioned at the base of the sting.

Note: If the end of the ventral hook is turned inward, it will rest on the sting shaft, and in holding it down will turn up the sting base for easier and proper positioning of the sting hook.

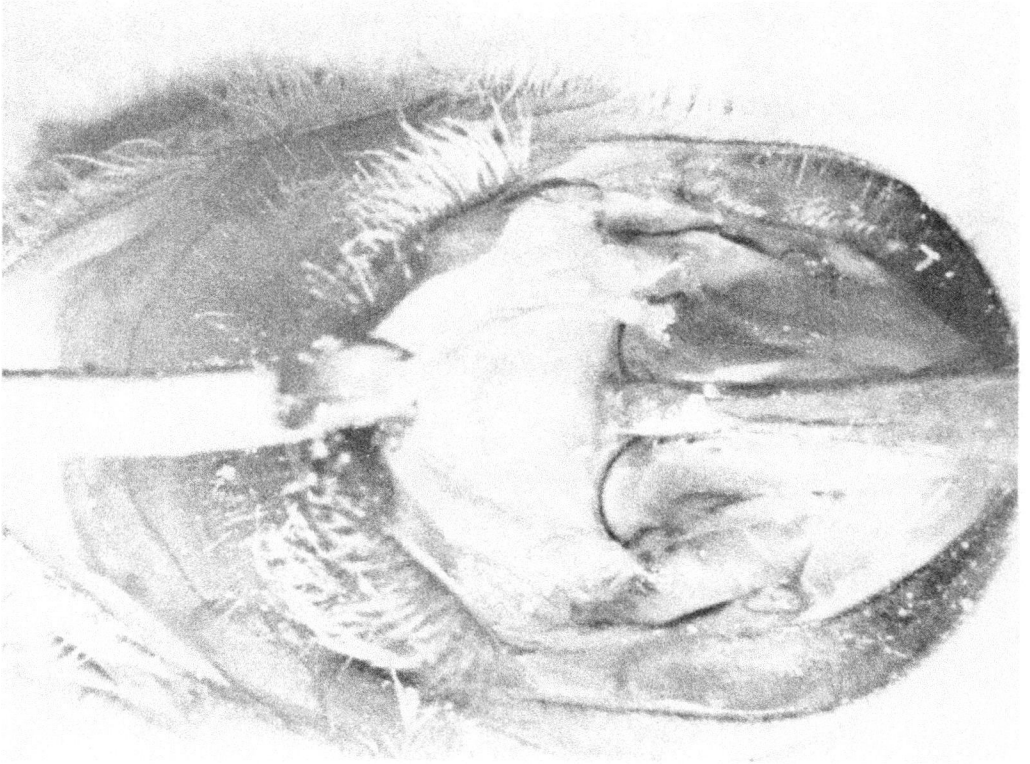

80. OPENING STING CHAMBER

The hooks are pulled outward to open the sting chamber and pull the sting base from over the vaginal orifice. The ventral edge of the vaginal orifice of the partly opened vagina is visible to the left of the sting hook as a "V" while the dorsal vaginal wall appears as a membrane immediately to the right of the "V."

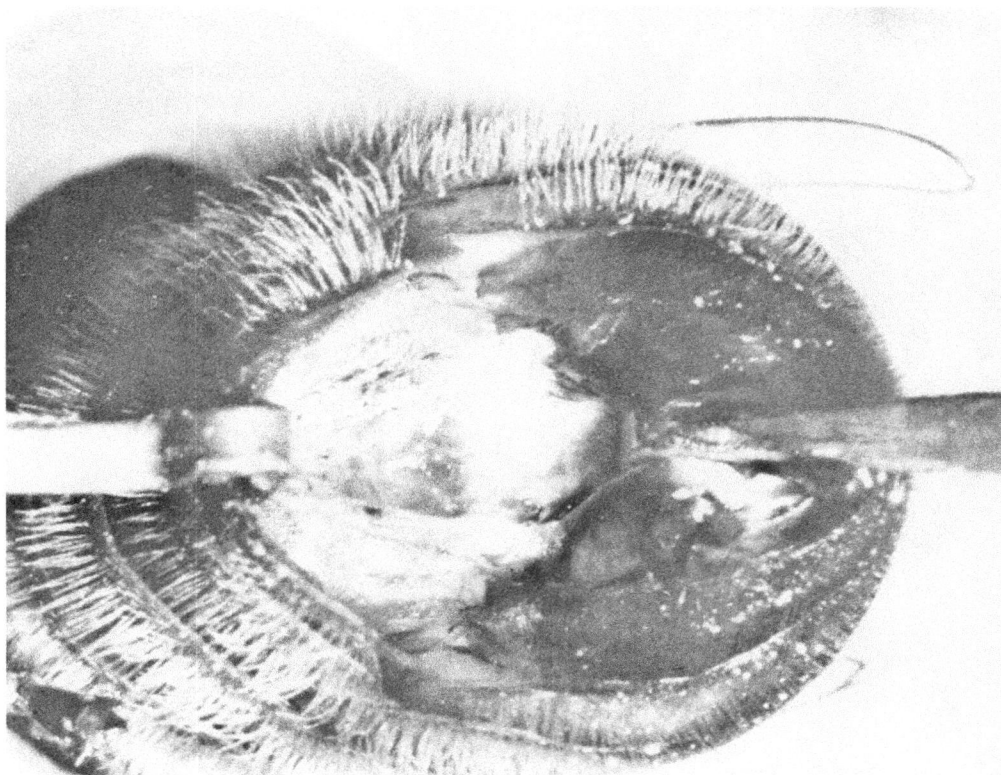

81. OPENING VAGINA FOR INSERTION OF THE SYRINGE

To avoid the necessity of lowering the valvefold with a probe or with the syringe tip, the vaginal orifice must be pulled partly or completely dorsal to the tip of the valvefold. Vaginal orifice stained.

Note: A drop of water in the sting chamber brings the vaginal orifice into sharp relief. For study only.

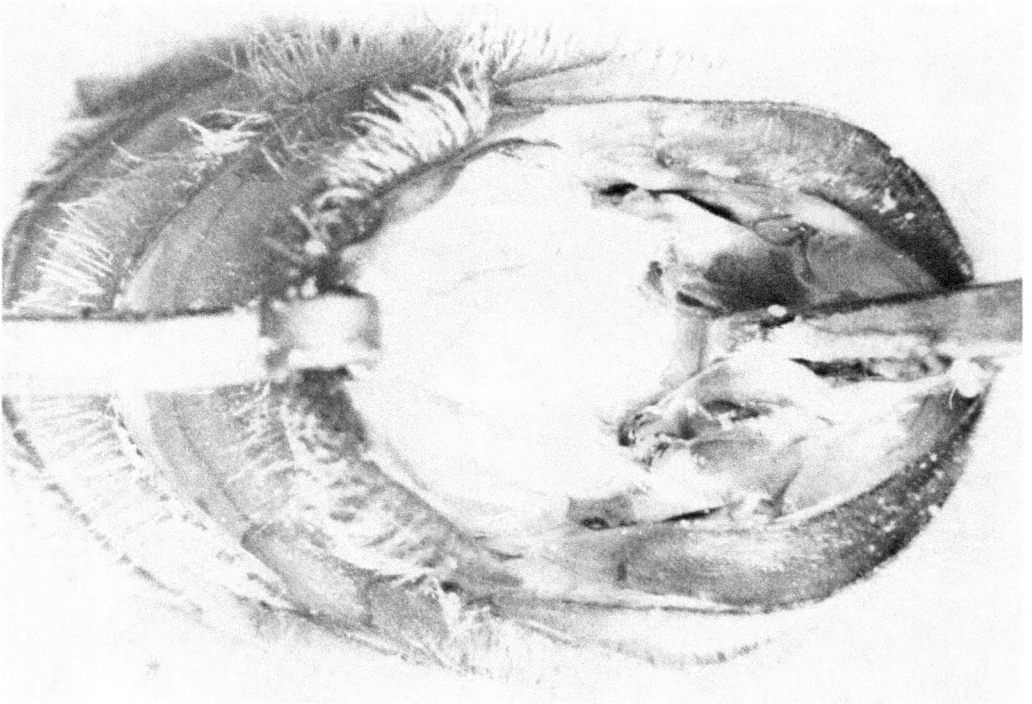

82. OPENING VAGINA FOR INSERTION OF THE SYRINGE

The outlines of the unstained vaginal orifice are less distinct but can be recognized. The inseminator works with unstained queens and must learn to distinguish the vaginal orifice.

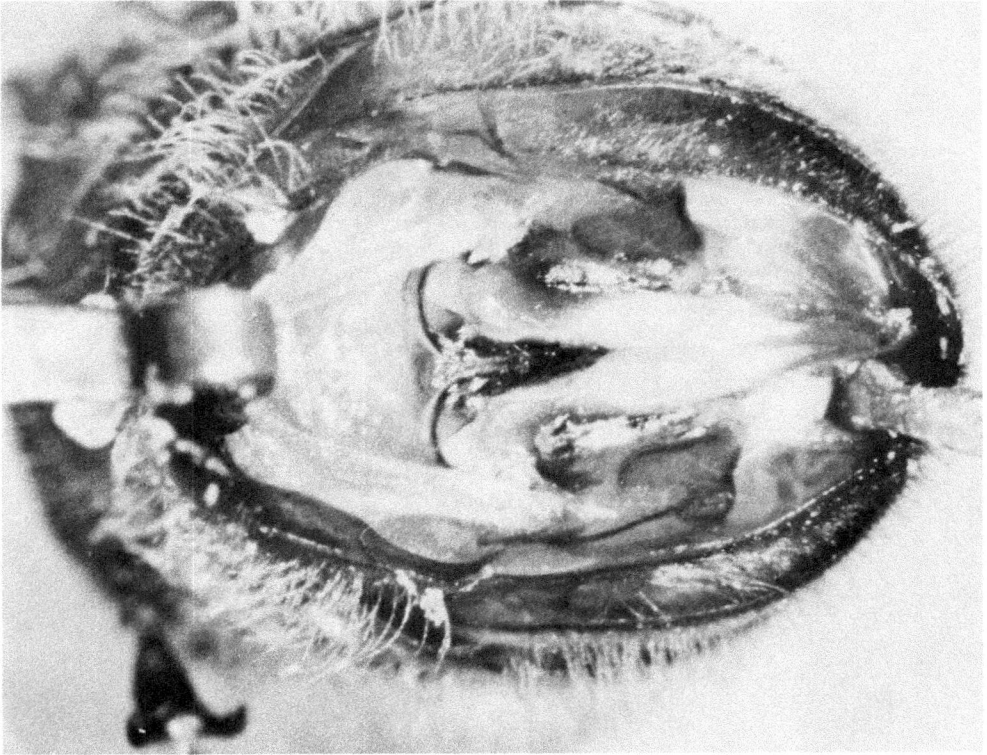

83. ALTERNATE METHOD OF POSITIONING STING HOOK

The sting chamber is opened by inserting the two hooks into the sting chamber and opening it to expose the sting base.

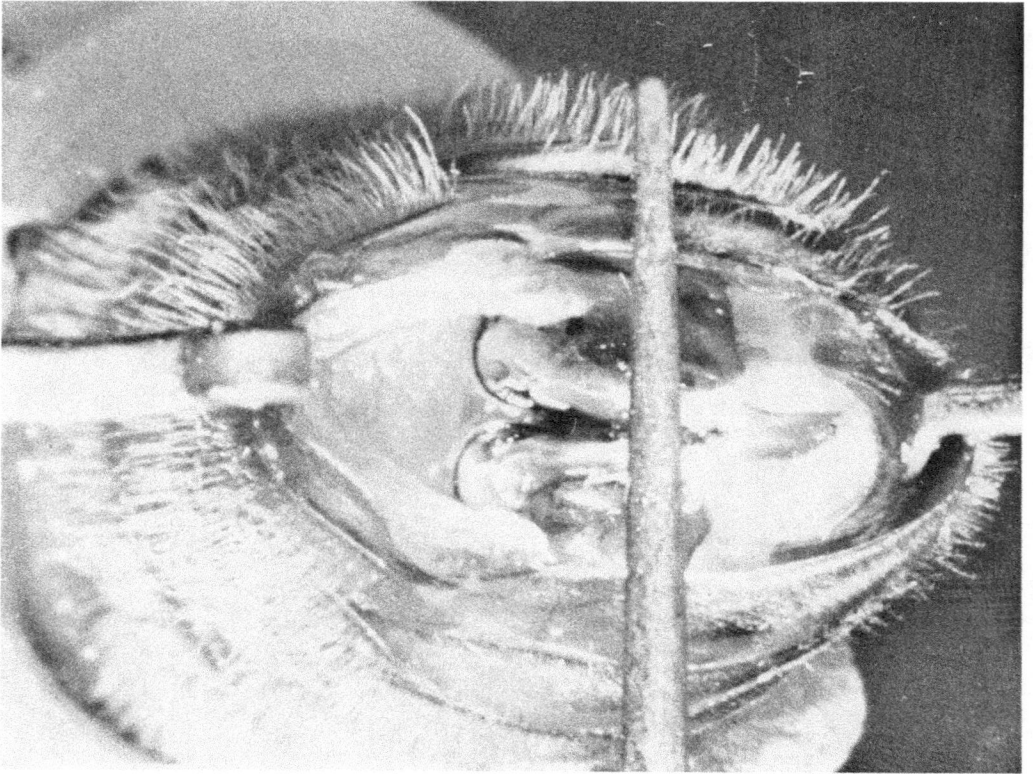

84. ALTERNATE METHOD OF POSITIONING STING HOOK

A pin is laid across the sting to hold it down while the sting hook is
moved toward the sting base.

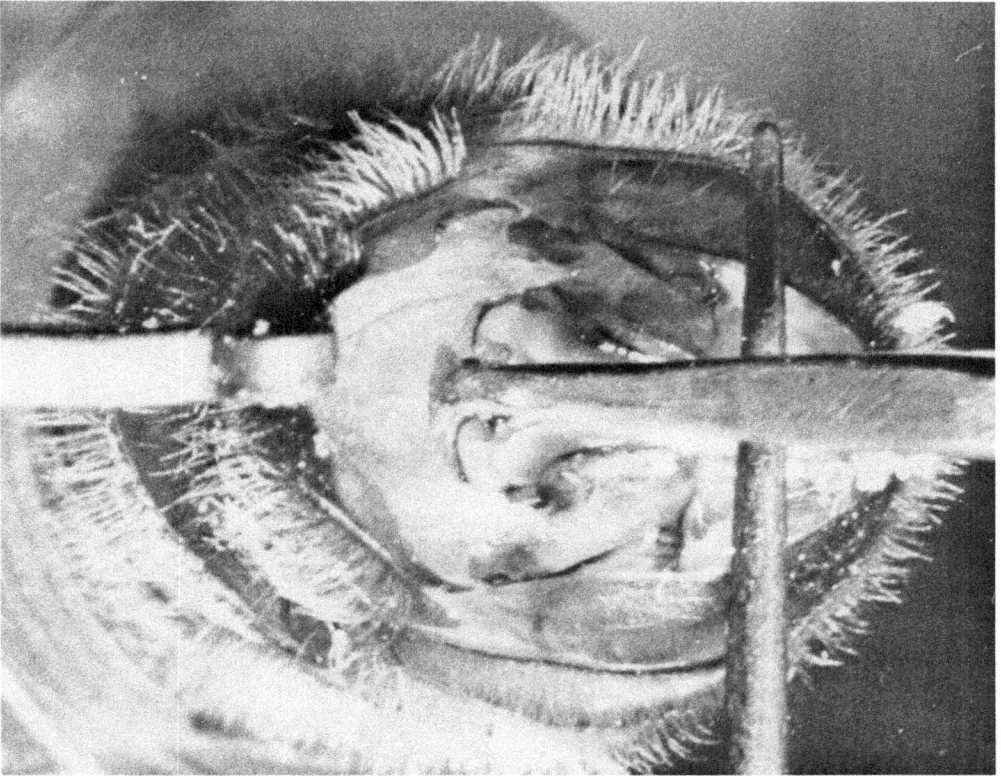

85. ALTERNATE METHOD OF POSITIONING STING HOOK

The sting hook is fitted against the base of the sting and the pin is removed. The sting can now be pulled dorsally until the vaginal orifice is dorsal to the tip of the valvefold.

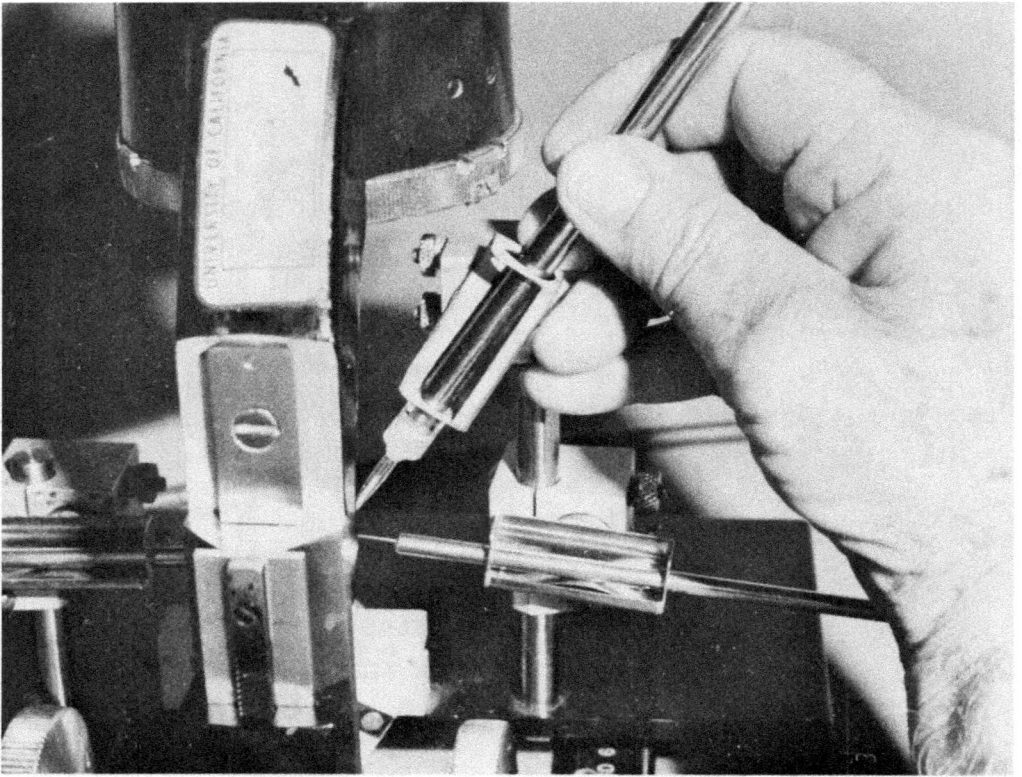

86. METHOD OF HOLDING SYRINGE

The syringe is fixed to the instrument by the same type of box that is used for the hooks. By grasping the syringe barrel between thumb and forefinger and steadying the 2nd and 3rd fingers against the end of the box the syringe may be slipped down or up in its holder with excellent control.

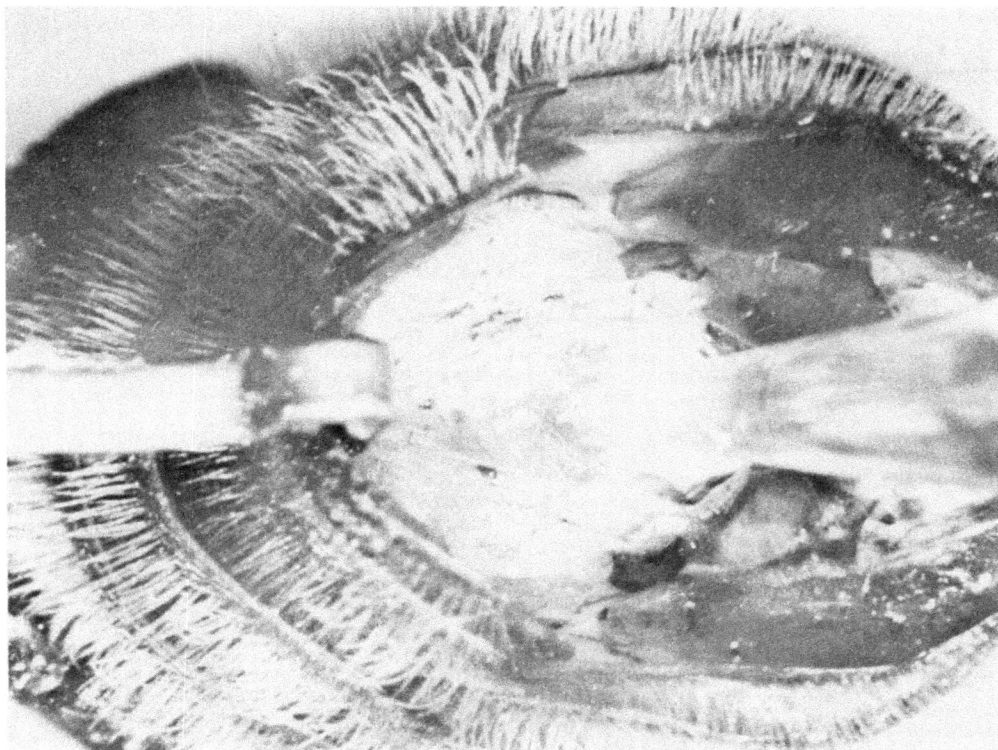

87. POSITIONING SYRINGE

The end of the syringe tip is positioned directly over the vaginal orifice dorsal to the "V" notch formed by the ventral edge. Vaginal orifice stained.

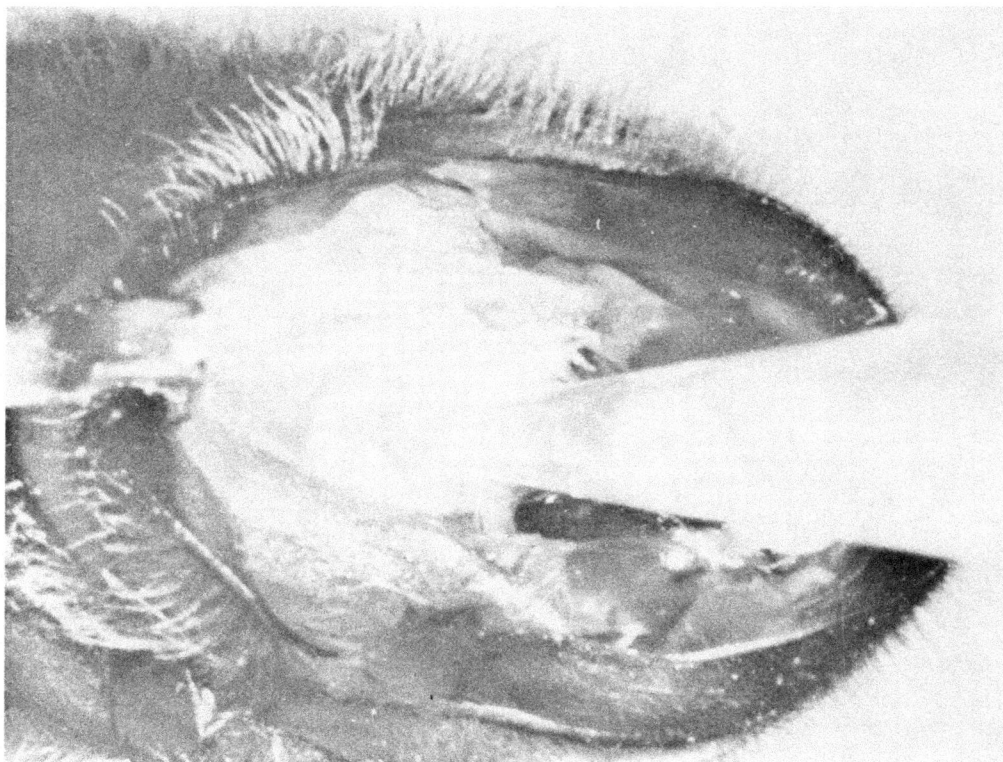

88. POSITIONING SYRINGE

The orifice of the unstained vagina is less distinct but can be recognized.

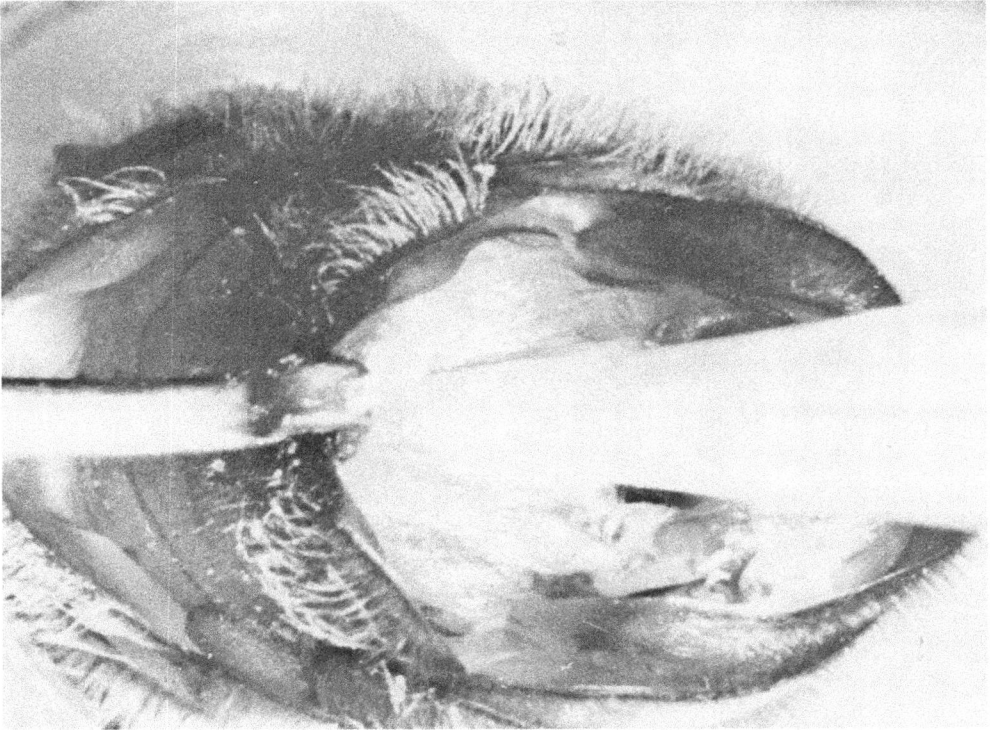

89. INSERTING SYRINGE

A slight amount of the plunger fluid, which was taken into the syringe following the semen load, is ejected to lubricate the tissues, and the syringe is inserted without lateral movement for about 1 mm into the vagina or until the tissues around the tip move. The syringe should **slip** easily into the vagina without movement of the surrounding tissues.

Note: Insertion of the syringe may be facilitated by a **slight** right and left rotation of the syringe as insertion proceeds.

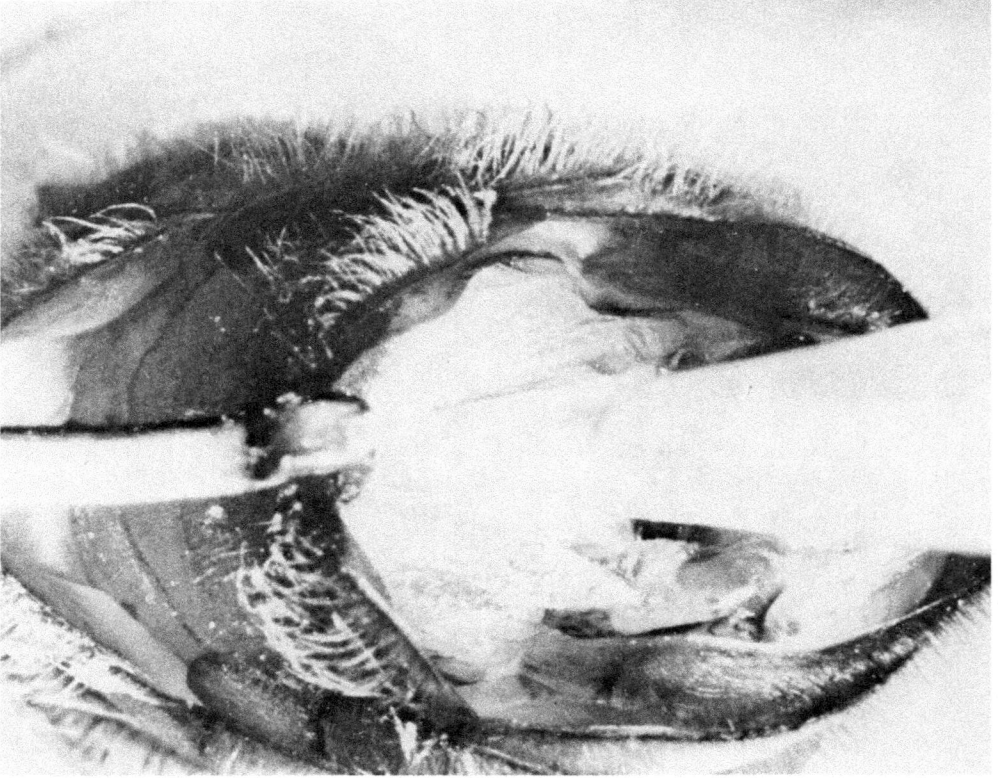

90. INJECTION OF SEMEN

If the semen moves easily down the tip as a **slight** clockwise test movement of the syringe plunger is made, ejection can proceed rapidly. If there is a little leakage, the syringe may be inserted slightly deeper. If the semen leaks in greater quantity, the syringe is improperly inserted or mucus was taken into the syringe and has occluded the median oviduct. Should the semen move only slightly in the syringe, semen may have dried to form a seal over the end of the tip.

Leaked semen may be removed by swabbing with a bit of facial tissue held with forceps. **Never** retake into the syringe any semen that has leaked, because semen that touched the sting chamber membranes may be contaminated.

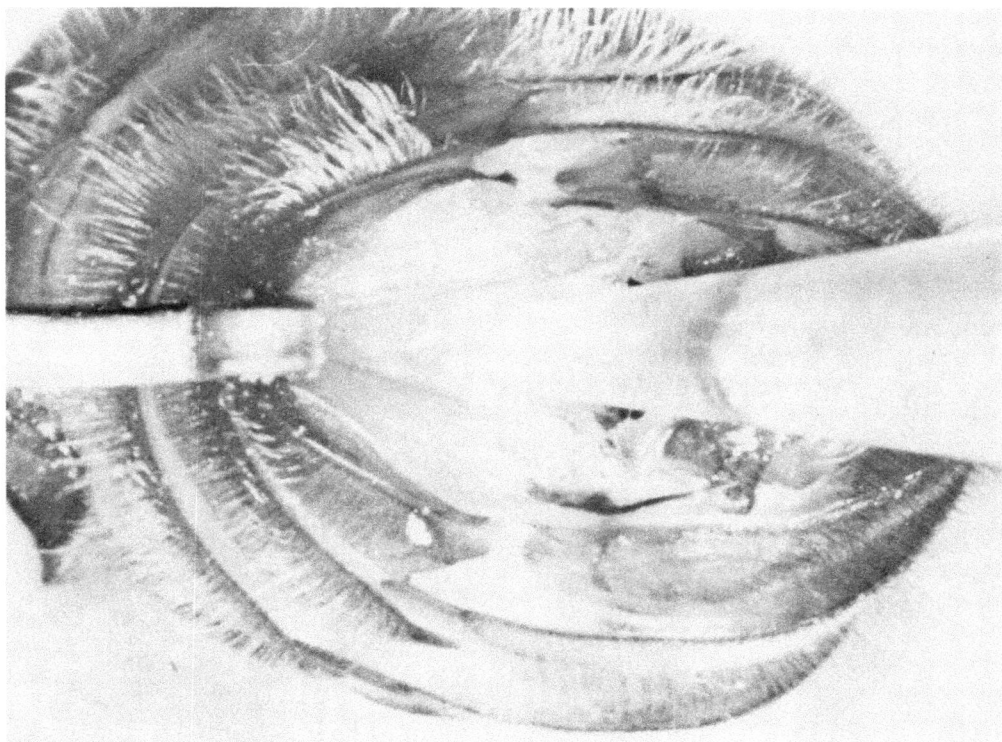

91. INJECTION OF SEMEN

All of the semen, and all or most of the air bubble should be injected.

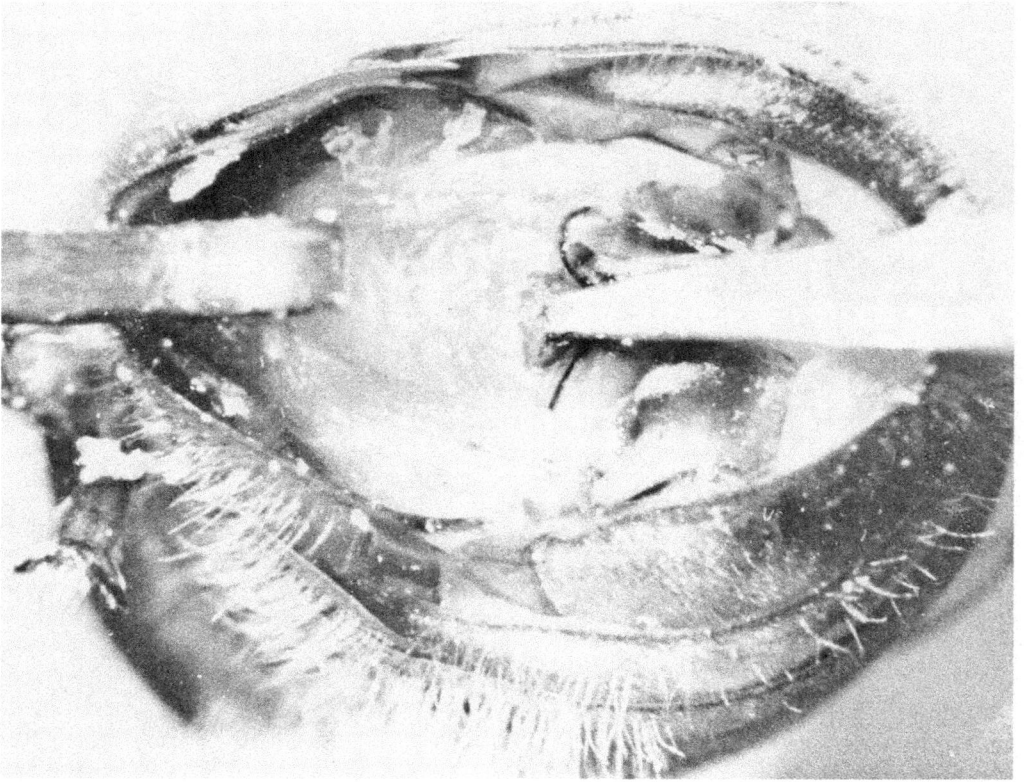

92. REMOVING HOOKS

The sting hook is disengaged and removed from the sting chamber.
The ventral hook is removed in a similar manner.

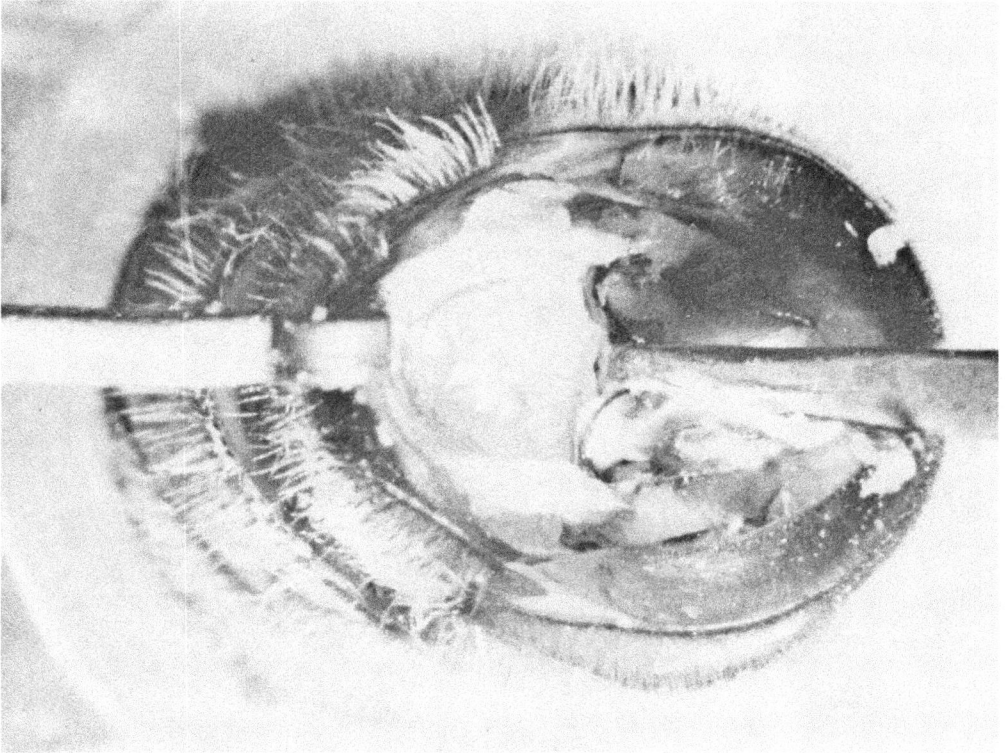

93. USE OF VALVEFOLD PROBE

The valvefold may be lowered with the probe supplied with the Mackensen instrument. The vagina is opened only slightly with the hooks. The orifice is not stretched.

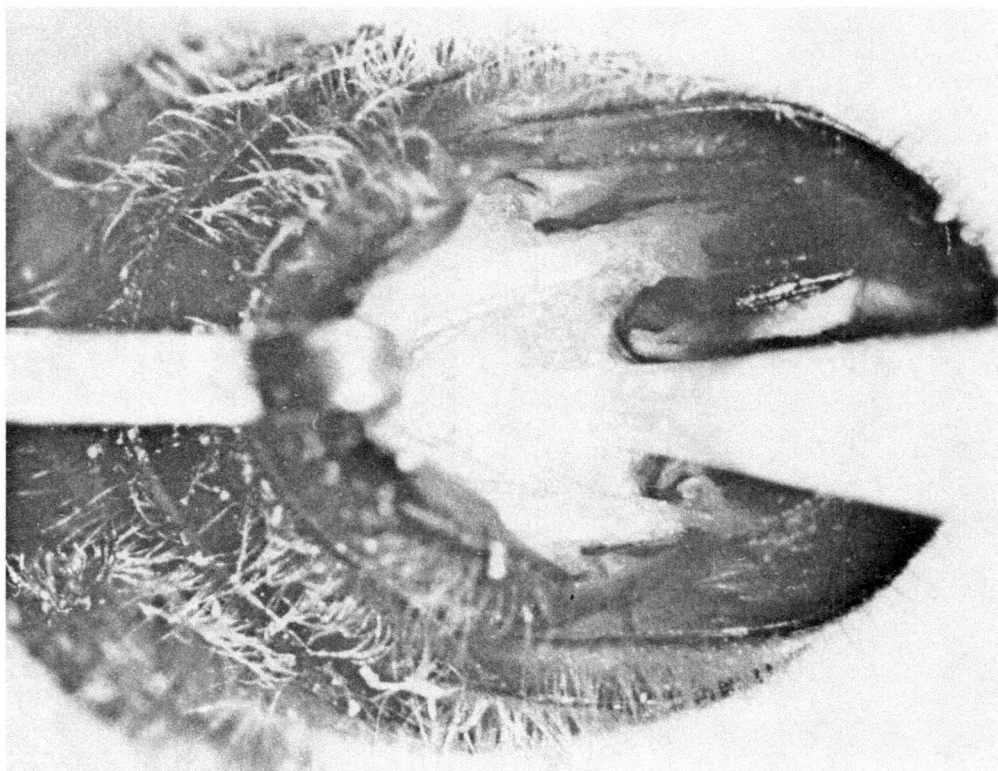

94. USE OF VALVEFOLD PROBE

The end of the syringe tip is positioned over the vaginal orifice or just ventral to it, and then is drawn upward while the valvefold is lowered.

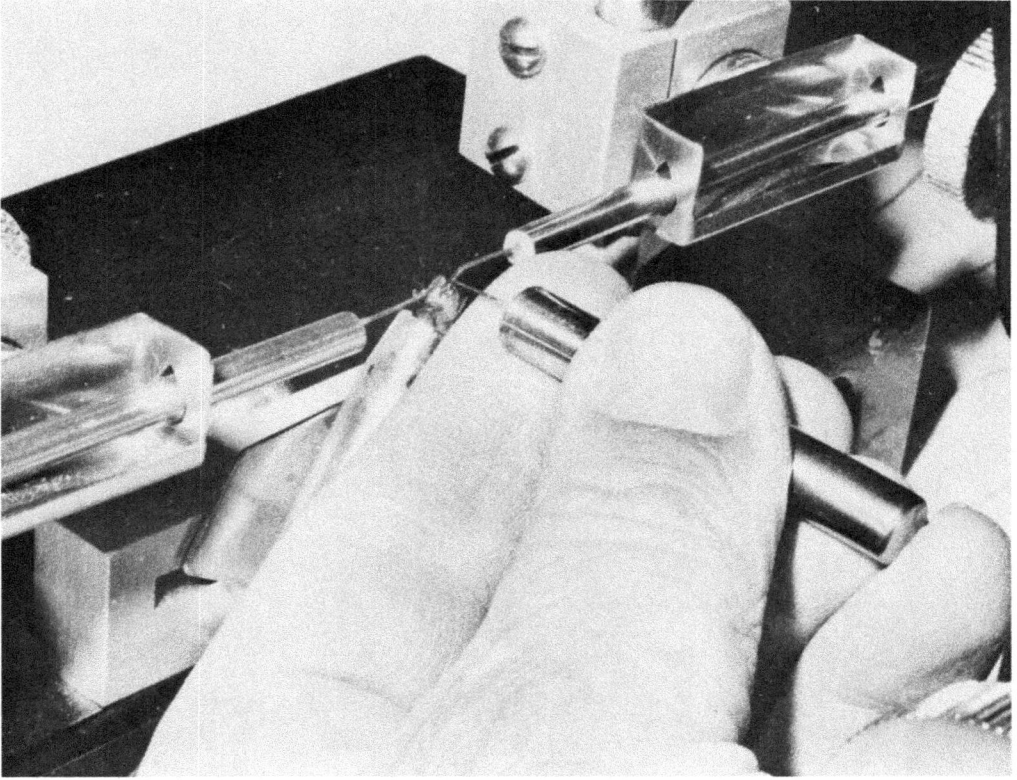

95. USE OF VALVEFOLD PROBE

Method of holding probe. The side of the forefinger rests lightly against the tube-holding block to steady the probe. The other fingers may rest on the microscope stage. The movement of the probe is controlled by slight movements of the thumb and forefinger.

96. USE OF VALVEFOLD PROBE

The end of the valvefold probe is positioned between the vaginal orifice and the sting and touches the membrane.

97. USE OF VALVEFOLD PROBE

The tip of the probe is pushed against the membrane and is moved toward the sternum. This opens the vagina and pushes the valvefold from over the median oviduct orifice.

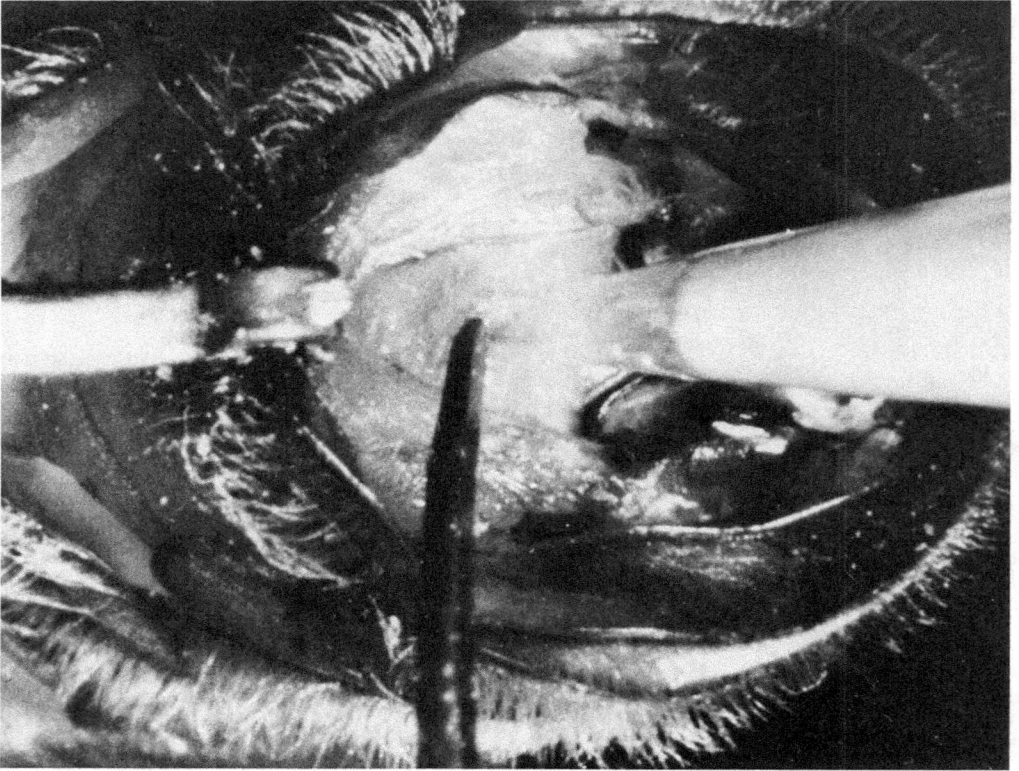

98. USE OF VALVEFOLD PROBE

The syringe is inserted just dorsal to the probe, and the probe is removed.

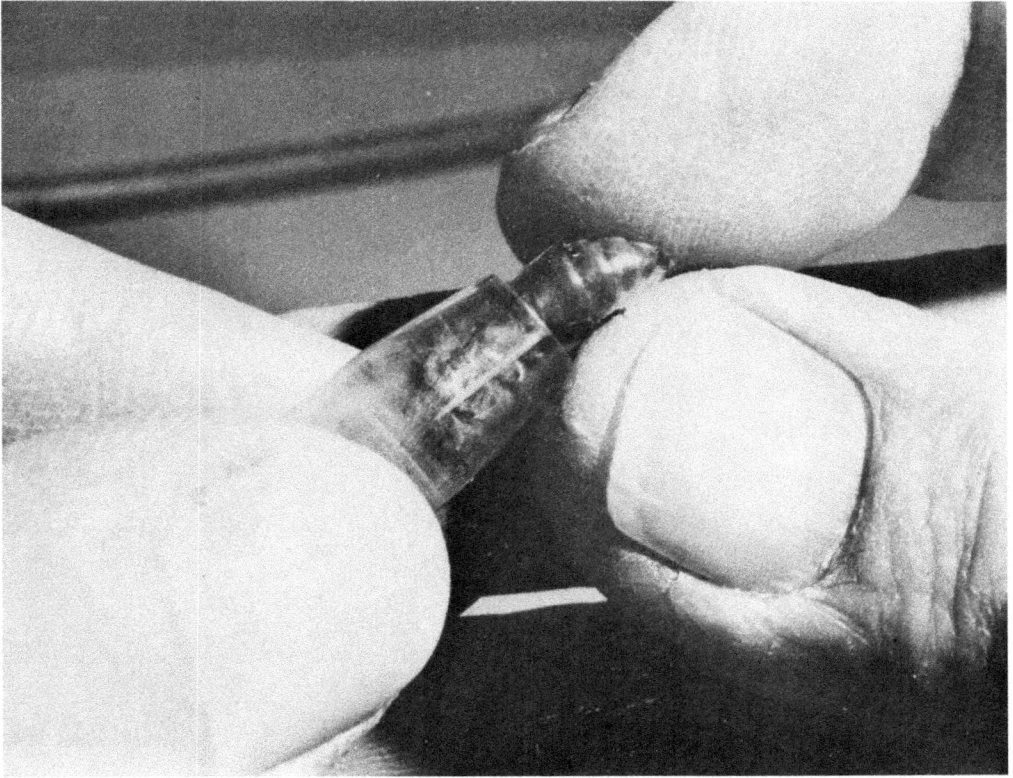

99. REMOVING QUEEN FROM TUBE

The queen is gently teased from the end of the tube.

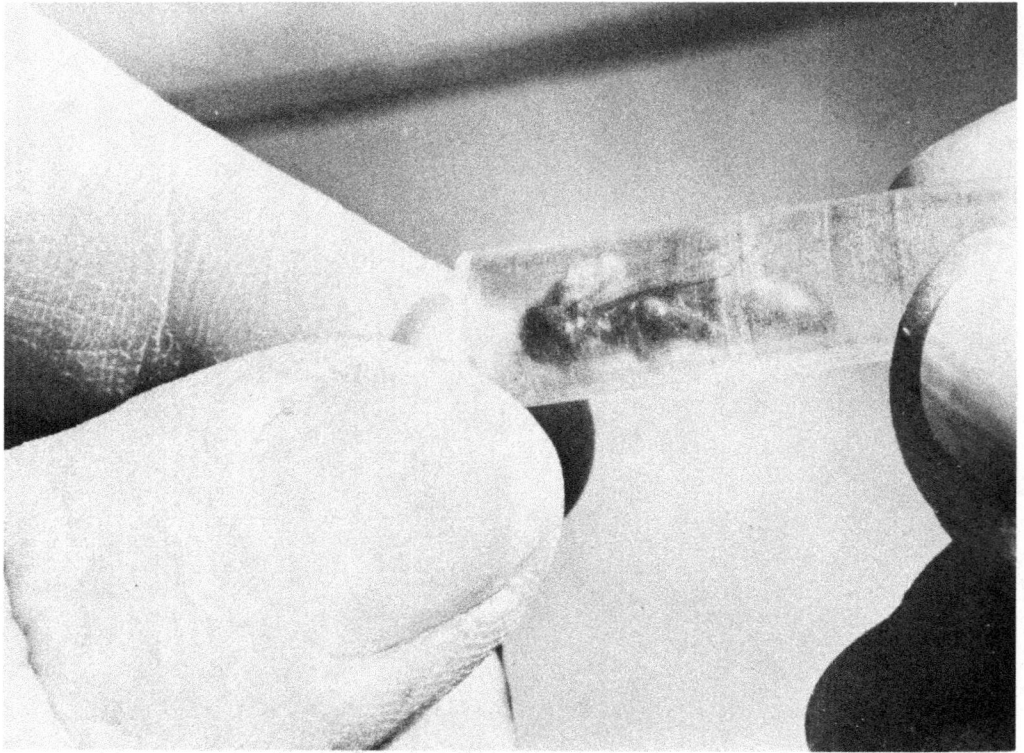

100. REMOVING QUEEN FROM TUBE

And is then blown into the cupped hand.

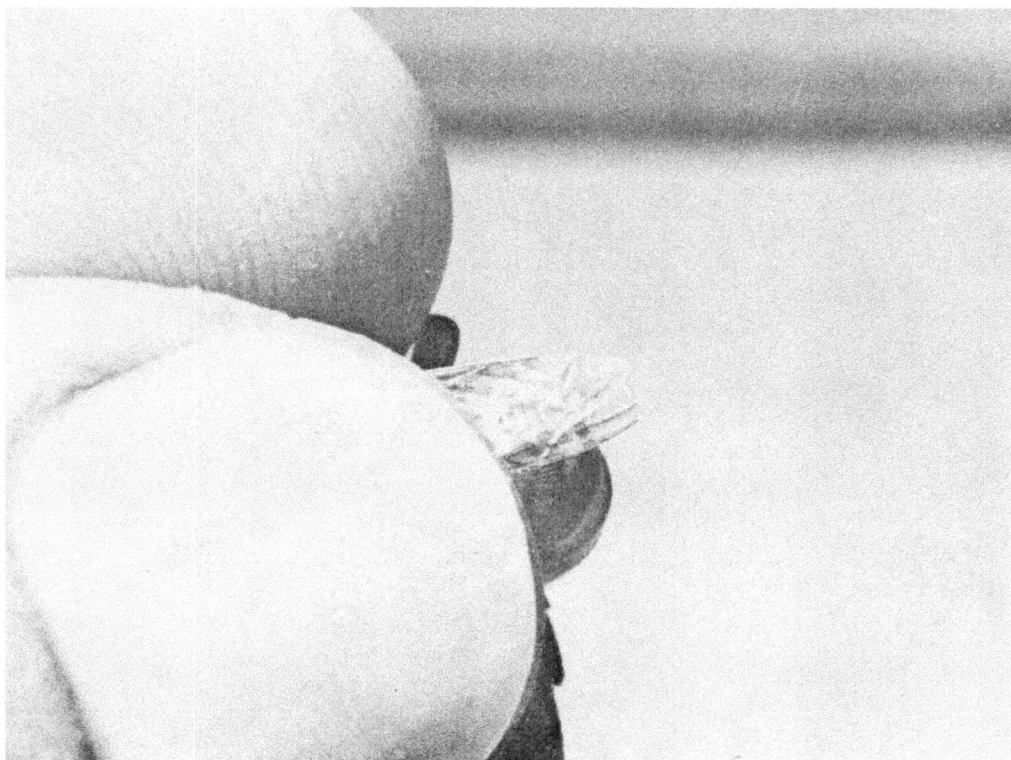

101. CLIPPING QUEEN

While not part of the insemination technique, a distinctive clip identifying artificially inseminated queens is useful. Clipping the tip of one wing serves this purpose.

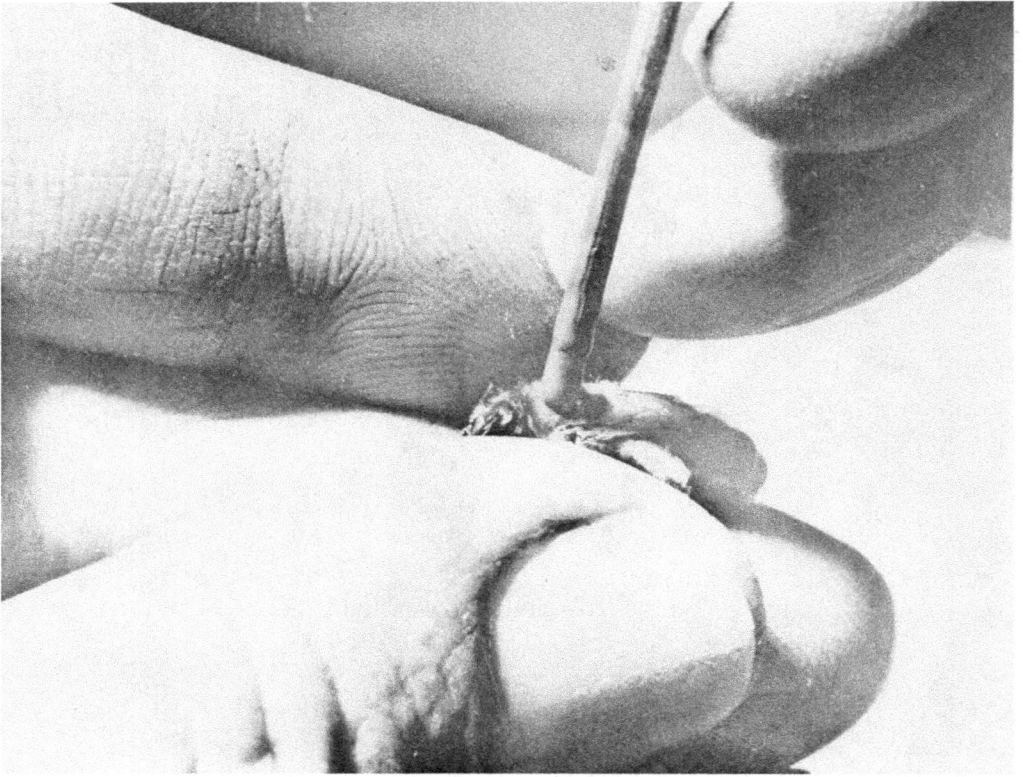

102. MARKING QUEEN

The immobile queens are easily marked following insemination, if desired.

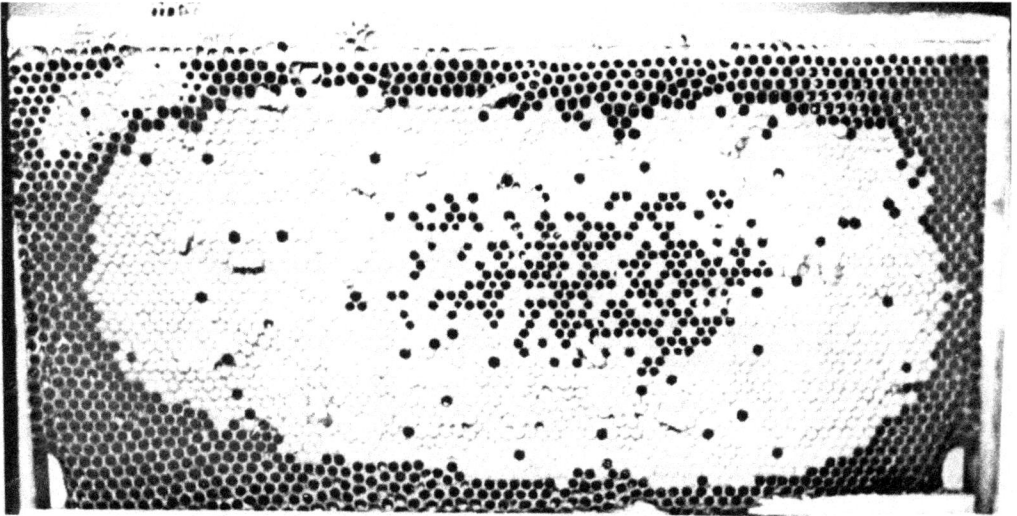

103. BROOD OF INSTRUMENTALLY INSEMINATED QUEEN

The performance of instrumentally inseminated queens is equal to that of their open mated sisters and varies according to stock and mating.

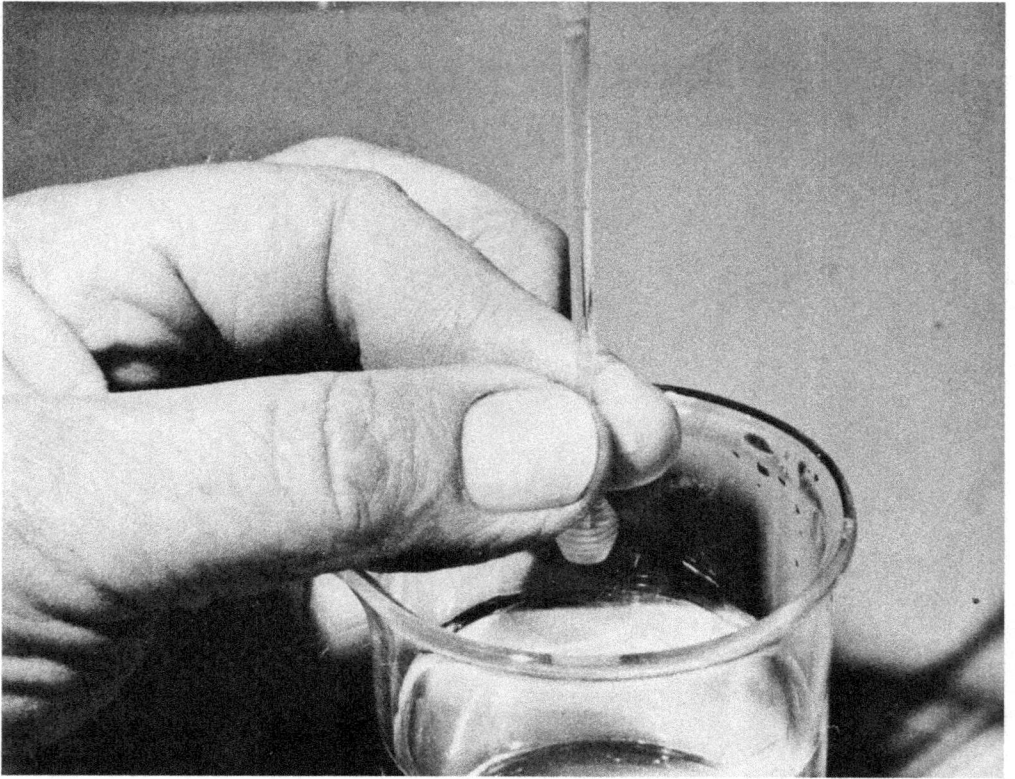

C. CARE OF SYRINGE

104. BACK WASHING TIP

Syringe tips are easily damaged or broken and should be handled carefully and kept clean. Cleaning the syringe between queens is not necessary unless the syringe has become contaminated or mucus has coagulated within the syringe cavity, or the drone stock is changed, but it must be cleaned before even a short period of non-use.

The plunger is retracted until it no longer touches the rubber diaphragm. The plastic tip is then removed from the adapter and is flushed by forcing water with a pipette through the tip from the distal end toward the base.

Note: A short piece of rubber tubing at the end of the pipette may protect the delicate end of the syringe tip from damage should the pipette be inclined during back washing.

105. BACK WASHING TIP

This is followed by dropping the tip, base down, into a beaker of sodium hypochlorite solution of 3% to 6% concentration for a period of several hours to overnight. The tip may be flushed with the solution if it is to be used immediately, but must be thoroughly flushed with water after sterilization.

Note: Do **not** use ethyl alcohol to sterilize plastic tips because this alcohol fractures the plastic.

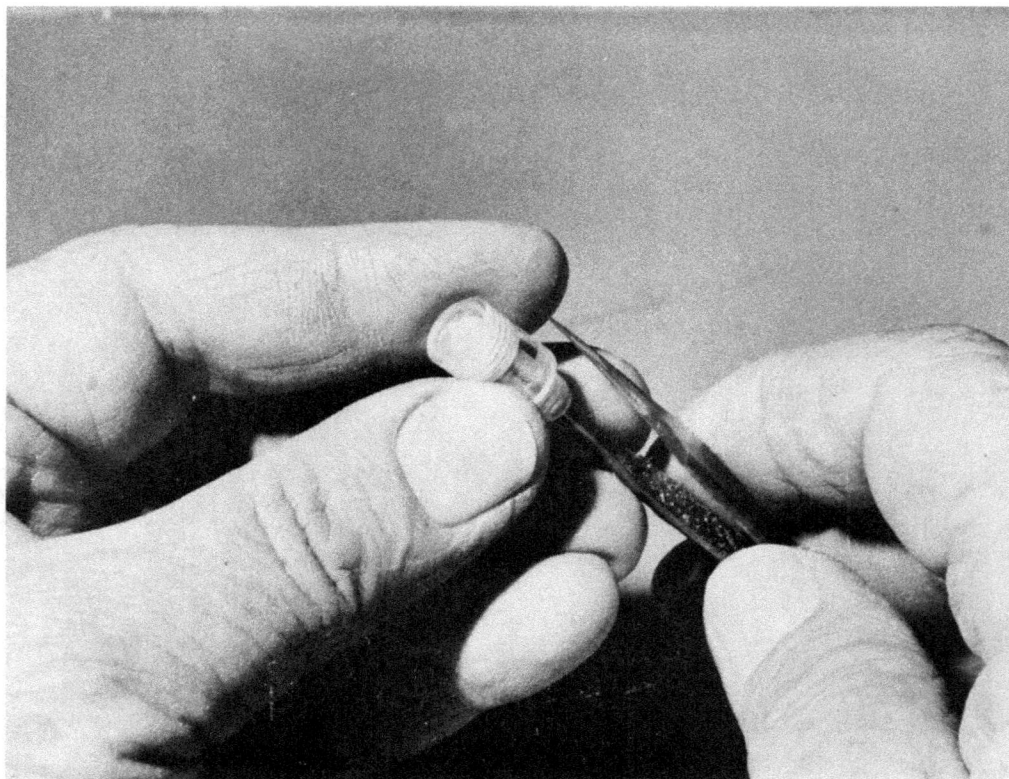

106. CLEANING ADAPTER AND DIAPHRAGM

The adapter is removed from the syringe barrel, and the diaphragm is dislodged with forceps. Both are dropped into the hypochlorite solution.

Note: Do not place metal in hypochlorite solution.

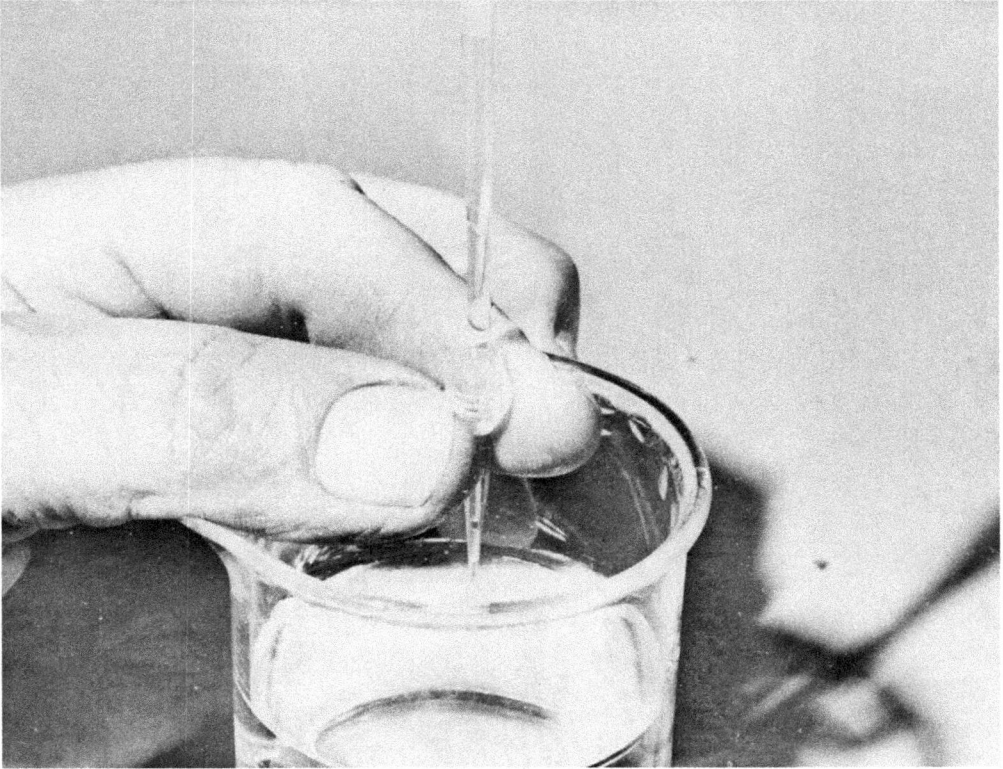

107. FLUSHING TIP

After sterilization in sodium hypochlorite, the tip, adapter, and diaphragm must be thoroughly rinsed and flushed before storage or use. The tip is rinsed under running water, or in clean water, and is flushed several times with clean water.

D. MAKING STING HOOK

Sting hooks are easily bent or broken. They can be made quickly from number 20 or 22 brass, bronze, or soft stainless steel wire.

108. FILING

The end of the wire is flattened and tapered with a fine file, or coarse stone.

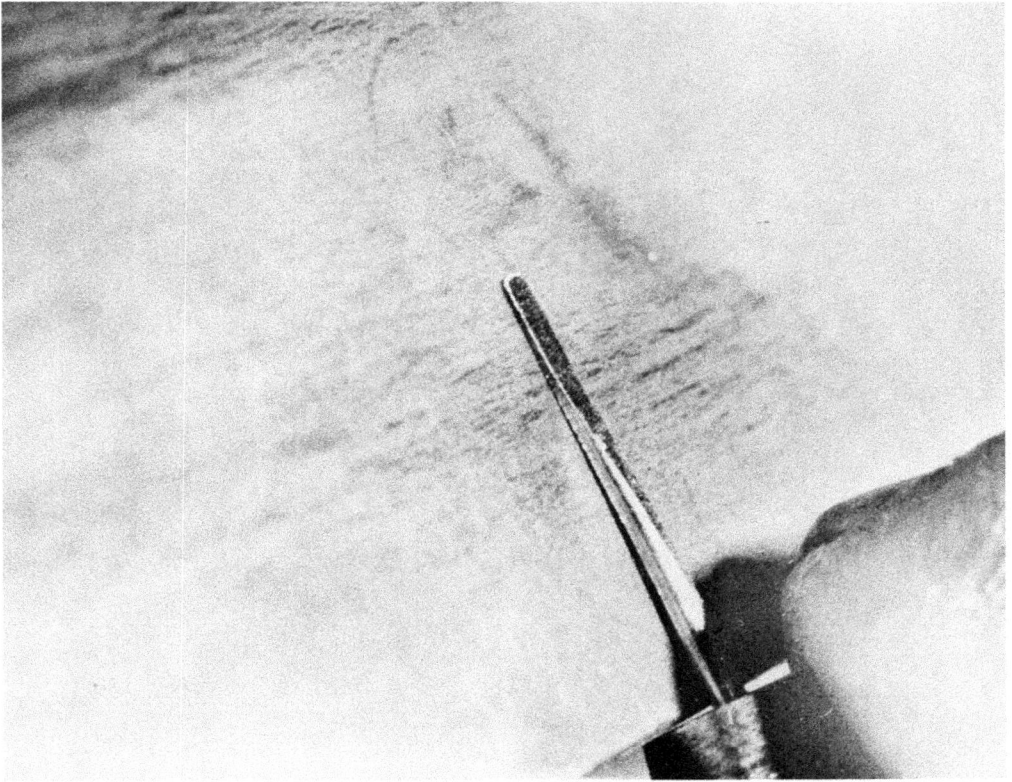

109. SMOOTHING ON STONE

The sides are smoothed and made parallel and the end thinned on a fine Arkansas stone.

110. NOTCHES

As a beginning in shaping the hook, notches are made in the edges of the flattened wire with the stone.

111. SHAPING

The hook is shaped on the stone, and the greatest width of the hook is made between 0.50 and 0.75 mm.

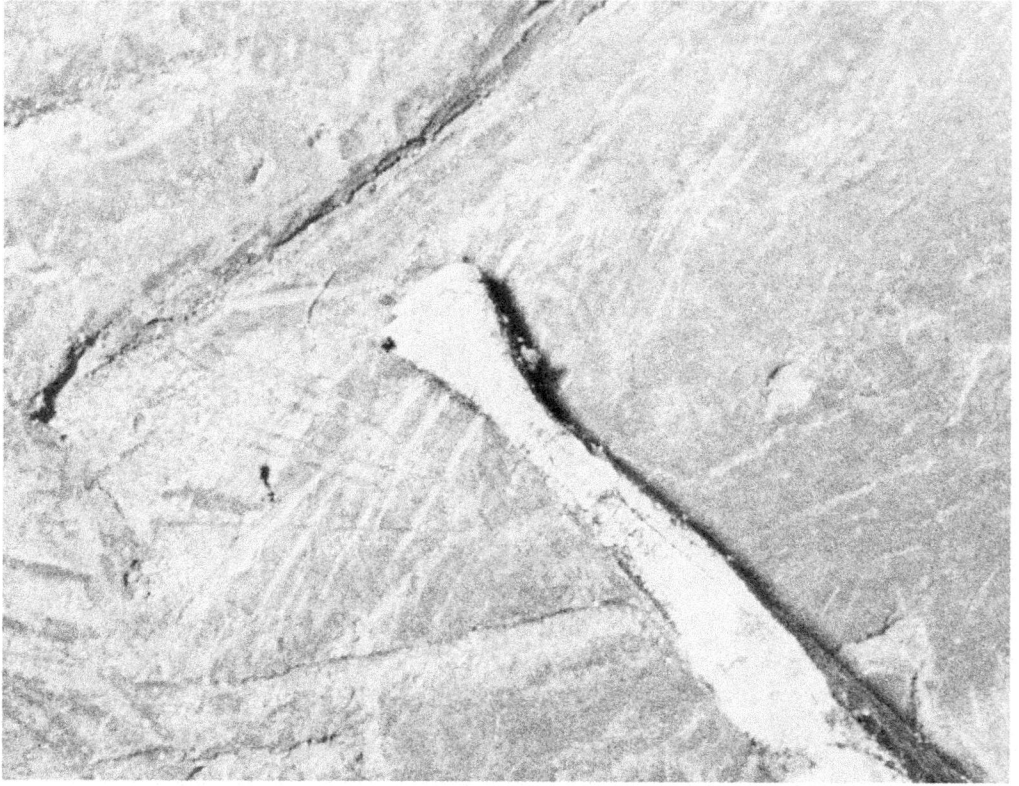

112. POLISHING

The shaped hook is polished with jeweler's rouge.

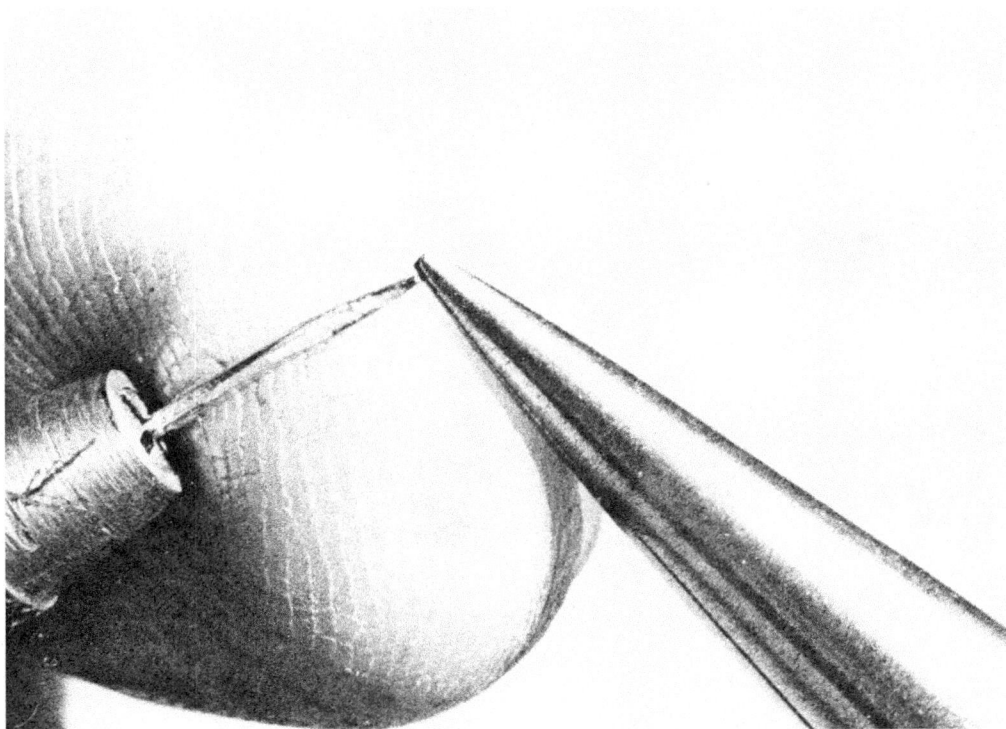

113. BENDING

The proper bend is made. It may vary to suit the instruments used and their adjustments.

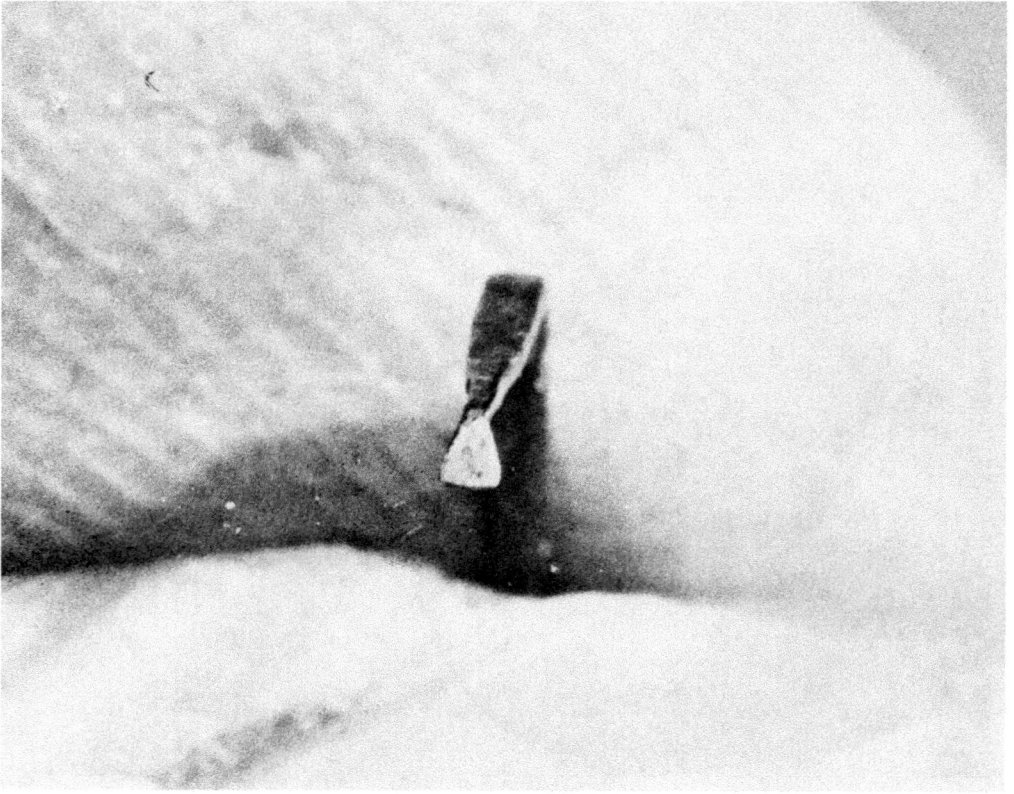

114. COMPLETED HOOK

E. INSTRUMENT CONSTRUCTION DIAGRAMS

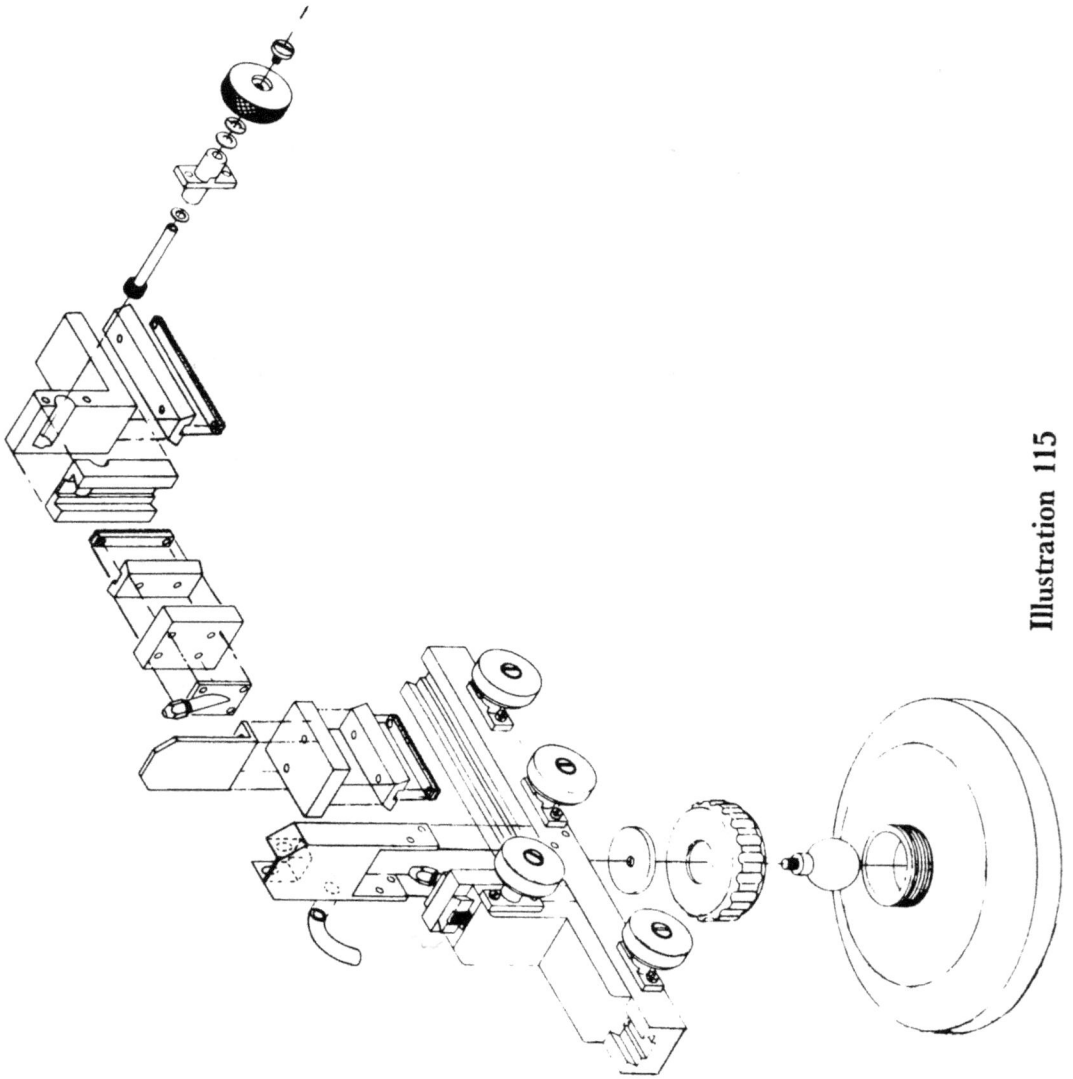

Illustration 115

E - 1 LAIDLAW INSTRUMENT

Exploded view of queen manipulator

The bar of the Laidlaw queen manipulator is 6 inches long. The maximum height of the manipulator is 3.25 inches from the lower surface of the base to the top of the anaesthetization chamber, and the weight of the manipulator is 3 pounds. This weight is great enough to provide stability, but not too great for easy lifting and moving over the microscope stage.

The length of the bar and the maximum height of the manipulator are important. The 6 inch bar provides for the necessary in and out movement of the pieces which bear the opening hooks, and allows the manipulator to be positioned on the microscope stage so the queen will be in the microscope field.

The 3.25 inch height is the maximum that will fit under many dissecting microscopes.

The queen and syringe manipulators are constructed of brass stock and the finished instruments are chrome plated.

Racks and pinions were standard mechanical stage items. Other, but similar, racks and pinions can be used instead.

Fluted edge

Bore $\frac{7}{16}$ deep.
Thread 24 TPI to
snug fit on ball socket

24 TPI
Press fit

Leather seat

Dot-Bolt

Drill and tap for $\frac{1}{4}$ - 28 NF

Drill and tap for 4-40
mach scr   6 Holes

Drill and tap for 4-40
mach scr   6 Holes

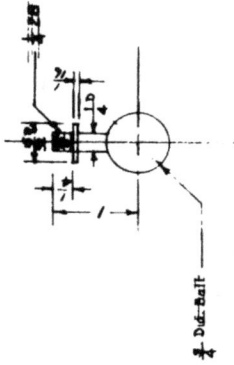

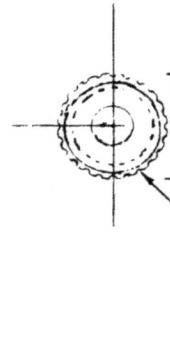

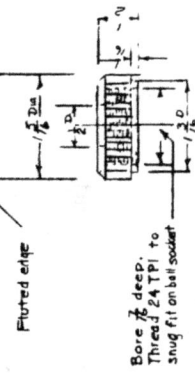

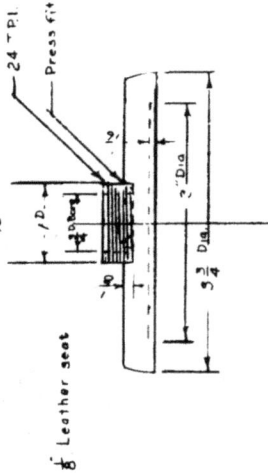

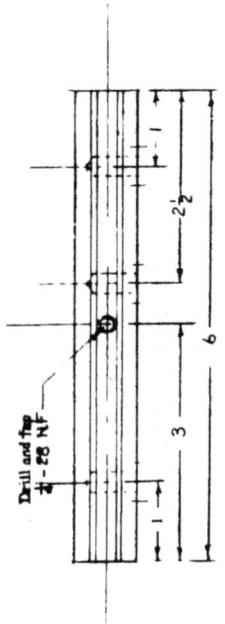

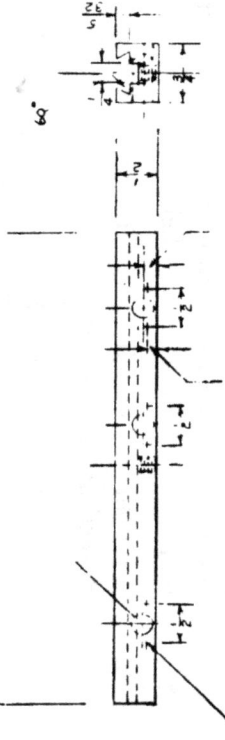

Note These dimensions
are determined by size
and type of rack and pinion
to insure proper mesh

Drill and tap 4-40   2 Holes

Clearance for
2-56 - 4 holes

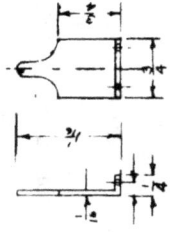

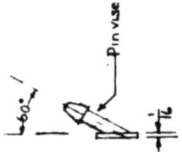

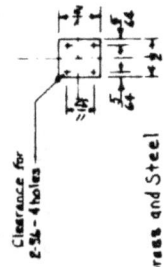

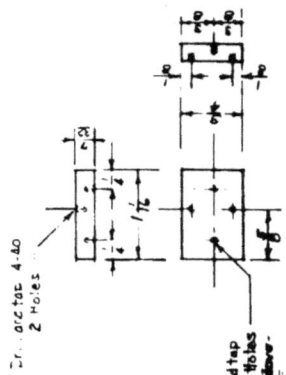

Drill and tap
4-40 2 holes
for std dove-
tail strip

Pin vise

Materials - Brass and Steel

Illustration 116

— 130 —

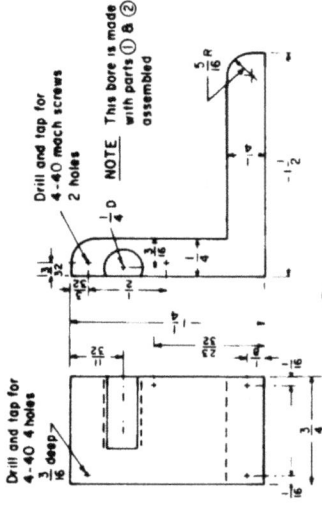

Drill and tap for 4-40 mach screws 2 holes

NOTE This bore is made with parts ① & ② assembled

Drill and tap for 4-40 4 holes 3/16 deep

②

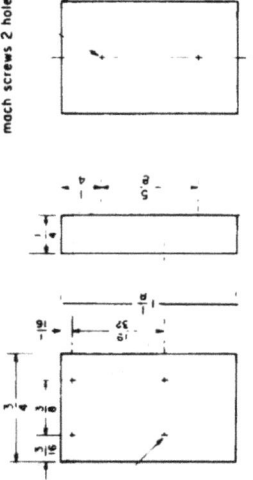

Drill and tap for 4-40 mach screws 2 holes

Drill and tap for 2-56 screws 4 holes

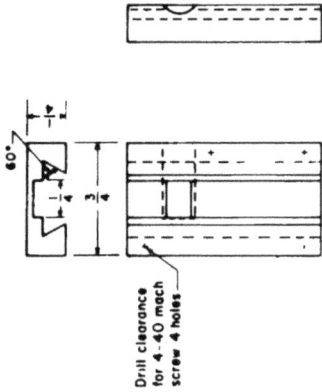

60°

Drill clearance for 4-40 mach screw 4 holes

①

Mill 1/8 slot 1/16 deep for spiral rack

Clearance holes for 4-40 screws
Ctr bore 2 holes

Drill and tap for 2-56 screws 2 holes

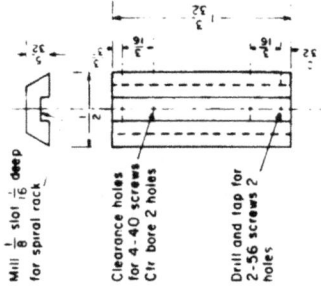

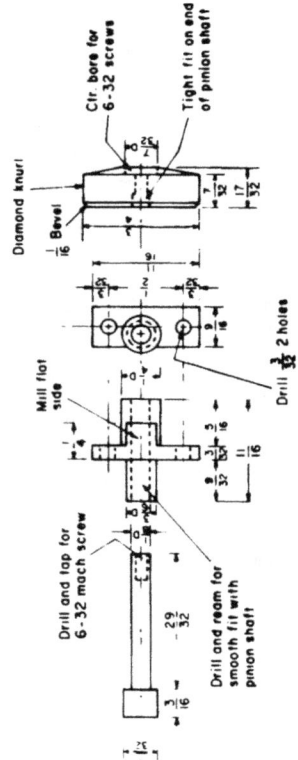

Diamond knurl

Bevel

Ctr bore for 6-32 screws

Tight fit on end of pinion shaft

Mill flat side

Drill 2 holes

Drill and tap for 6-32 mach screw

Drill and ream for smooth fit with pinion shaft

Illustration 117

— 131 —

Illustration 118

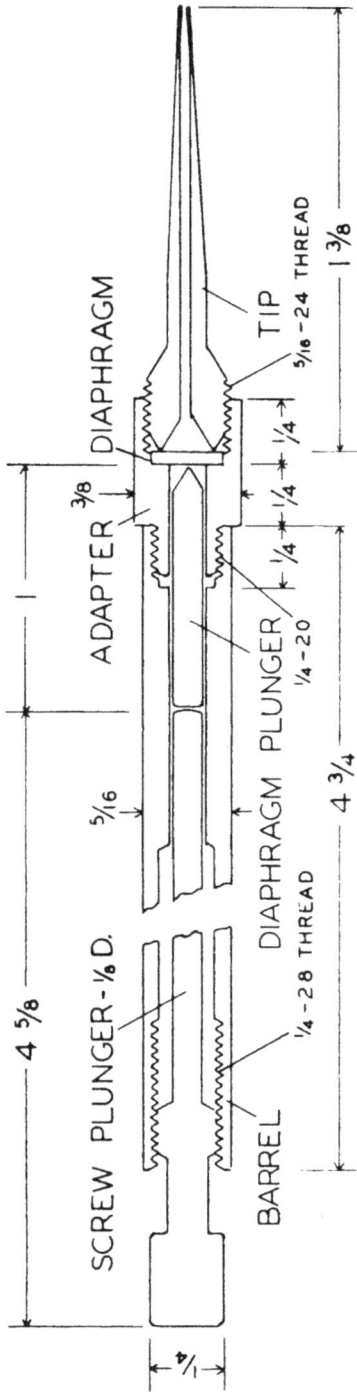

SCREW PLUNGER - ⅛ D.

DIAPHRAGM PLUNGER

BARREL

¼ - 28 THREAD

ADAPTER ³⁄₈

DIAPHRAGM

TIP

⅝⁄₁₆ - 24 THREAD

¼ - 20

4 ⅝

5⁄16

4 ¾

1 ³⁄₈

¼ ¼ ¼ ¼

³⁄₈

¼

SYRINGE

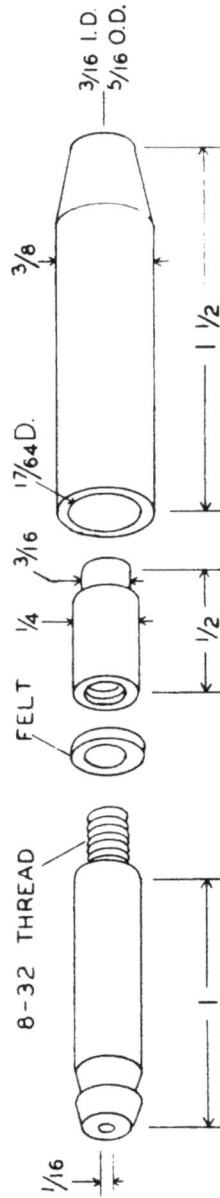

3⁄16 I.D.
5⁄16 O.D.

3⁄8

1 ½

17⁄64 D.

3⁄16

¼

FELT

8-32 THREAD

½

½

1

1⁄16

QUEEN HOLDER

Illustration 119

E - 2 MACKENSEN INSTRUMENT (from USDA Agricultural Handbook No. 390)
See this handbook and **USDA Manual ET-250** for details of construction.

— 133 —

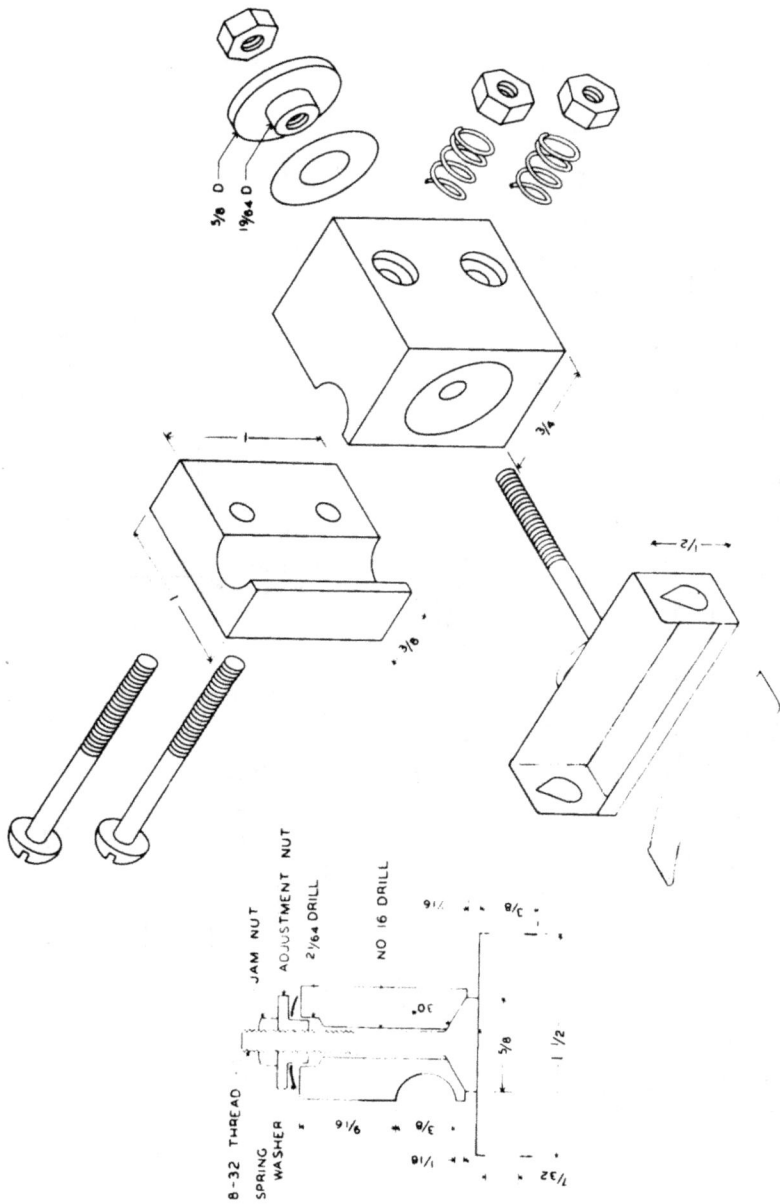

Illustration 120

5/8 D
19/64 D

8-32 THREAD
SPRING WASHER
JAM NUT
ADJUSTMENT NUT
21/64 DRILL
NO 16 DRILL

3/4

1

3/8

1/2

30°

5/8

1 1/2

7/16

3/8

1

9/16

3/8

1/16

7/32

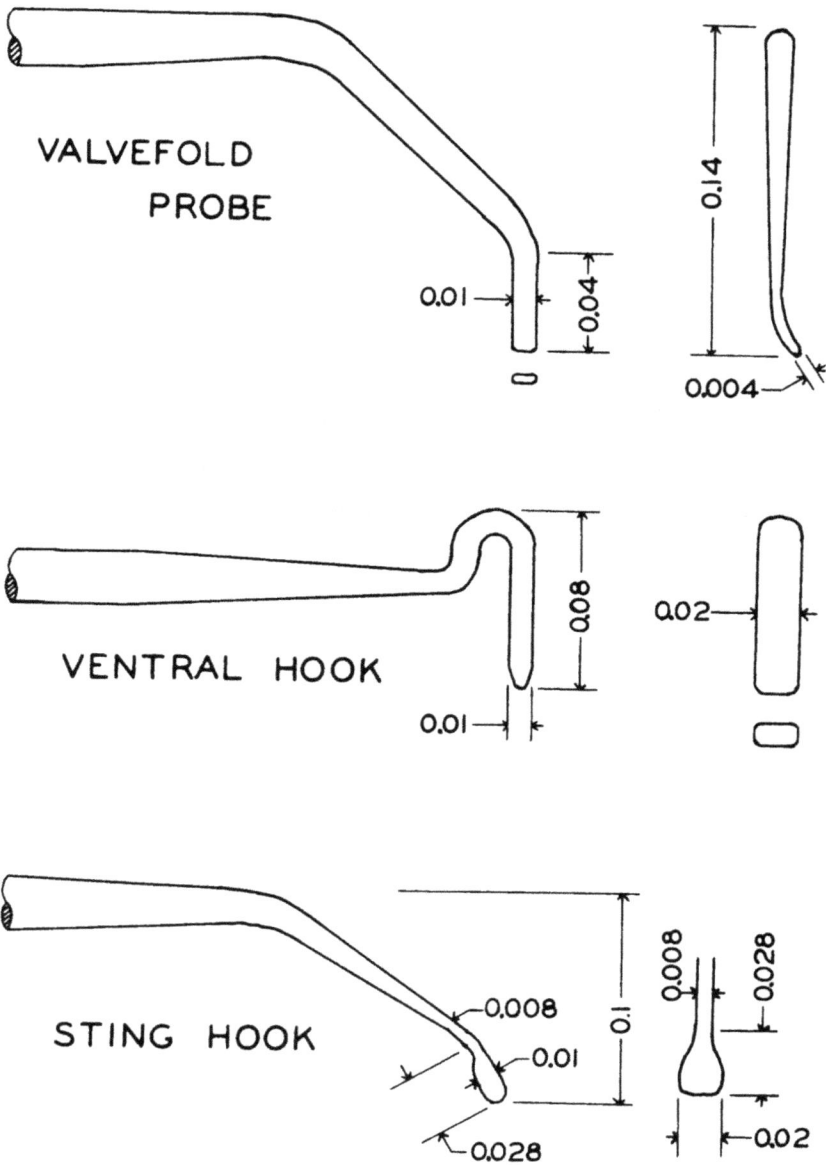

VALVEFOLD PROBE

0.01

0.04

0.14

0.004

VENTRAL HOOK

0.08

0.01

0.02

STING HOOK

0.008

0.01

0.1

0.028

0.008

0.028

0.02

Illustration 121

E - 3 RUTTNER - SCHNEIDER - FRESNAYE INSTRUMENT

See Ruttner 1975. **Die Instrumentelle Besamung der Bienenkonigin** for details of construction. (Photographs courtesy of F. Ruttner.)

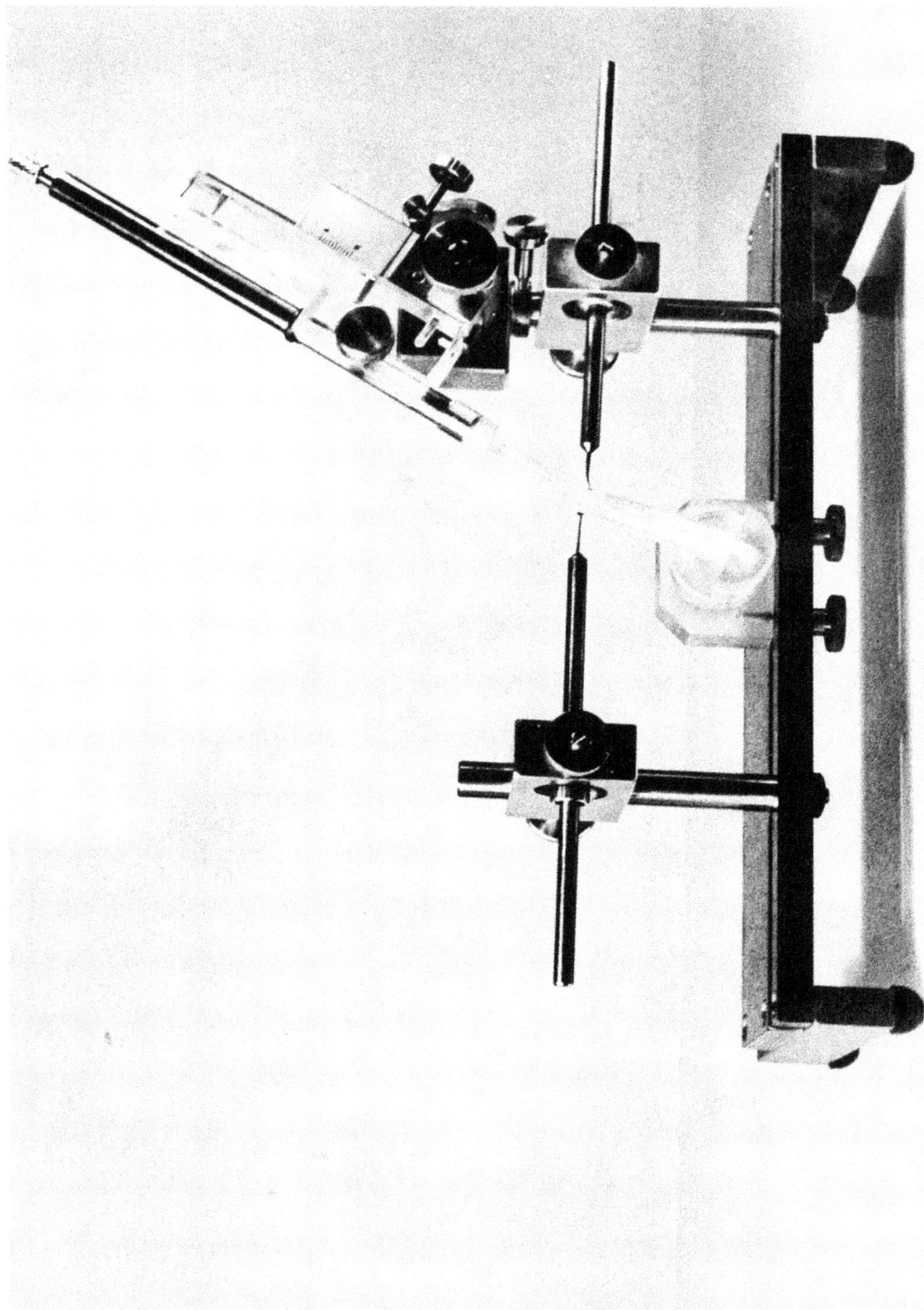

Illustration 122

Illustration 123

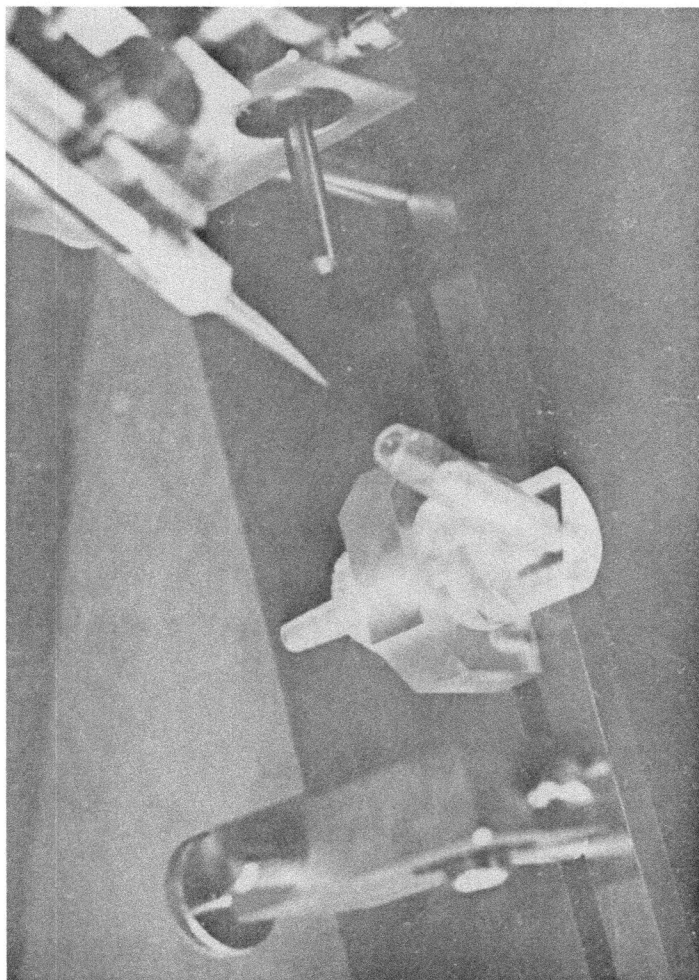

Illustration 124

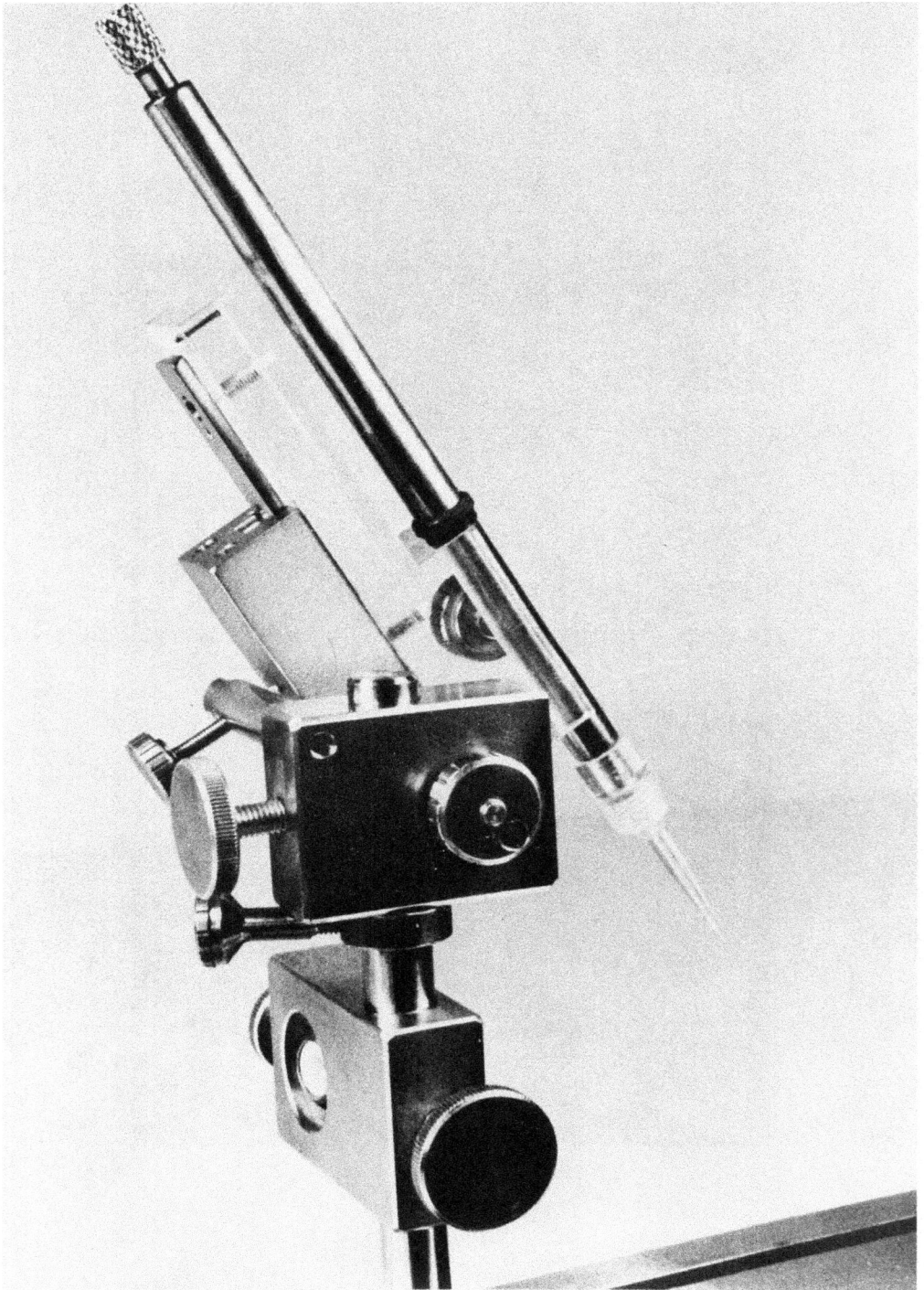

Illustration 125

F. SUGGESTIONS FOR FURTHER INFORMATION

Bishop, G. H. 1920. Fertilization in the honeybee. I. The male sexual organs — their histological structure and physiological functioning. Jour. Expt. Zool. 31(2):225-258. II. Disposal of the sexual fluids in the organs of the female. 31(2):267-286.

Breslau, Ernst von. 1906. Der Samenblasengang der Bienenköningen. Zool. Anz. 29(10):299-323.

Camargo, C. A. de. 1972. Aspectos da reprodução dos Apideos sociais. Dissertação apresentada a Faculdade de Medicina de Ribeirão Preto (USP) para obtenção do Grau de Mestre em Ciencias. 63 pp.

————. 1975. Métodos de contrôle de fecundação natural e instrumental. In: Anais do 3rd Congresso Brasileiro de Apicultura. pp. 131-136.

Camargo, Joao, M. F. 1972. Técnicas de contrôle de cruzamentos. In: Camargõ, Joao Maria Franco de (editor). Manual de Apicultura, pp. 59-96. Editôra Agronômica "Ceres" LTDA. São Paulo.

Camargo, J. M. F., and L. S. Goncalves. 1968. Note on techniques for instrumental insemination of queen honeybees. J. Apic. Res. 7(3):157-161.

————. 1971. Manipulation procedures in the technique of instrumental insemination of queen honeybee *Apis mellifera* L. (Hymenoptera: Apidae). Apidologie 2(3):239-246.

Camargo, J. M. F. and M. L. S. Mello. 1970. Anatomy and histology of the genital tract, spermatheca, spermathecal duct and glands of *Apis mellifera* queens (Hymenoptera: Apidae). Apidologie 1(4): 351-373.

Chaud-Netto, José. 1975. Aspectos do melhoramento da *Apis mellifera*. In: Anais do 3rd Congresso Brasileiro de Apicultura. pp. 77-91.

Huber, Francis. 1814. New Observations Upon Bees. Translation from the French by C. P. Dadant 1926. American Bee Journal, Hamilton, Illinois. 230 pp.

Laidlaw, H. H. Jr. 1932. Hand mating of the queen bee. Amer. Bee J. 72(7):286.

————. 1934. The Reproductive Organs of the Queenbee in Relation to Artificial Insemination. Thesis submitted to the Faculty of the Louisiana State University and Agricultural and Mechanical College in partial fulfillment of the requirements for the degree of Master of Science.

————. 1939. The Morphological Basis for an Improved Technique of Artificial Insemination of Queenbees of *Apis mellifica Linnaeus*. A thesis submitted to the Graduate School of the University of Wisconsin in partial fulfillment of the requirements for the degree of Doctor of Philosophy.

. 1944. Artificial insemination of the queen bee (*Apis mellifera* L.) J. Morph. **74**(3):429-465.

. 1949a. New instruments for artificial insemination of queen bees. Amer. Bee J. **89**(12):566-567.

. 1949b. Development of precision instruments for artificial insemination of queen bees. J. Econ. Ent. **42**(2):254-261.

. 1953. An anaesthetization chamber for the artificial insemination of queen bees. J. Econ. Ent. **46**(1):167-168.

. 1956 (1958). Organization and operation of a bee breeding program. **In:** Proceedings Tenth International Congress of Entomology, **4**:1067-1078.

Laidlaw, H. H. Jr. 1976. Instrumental Insemination of Queen Honey Bees. Slide Set. Dadant and Sons, Inc., Hamilton, Illinois.

Laidlaw, Harry H. and Coby Lorenzen. 1957. Microsyringe adapter. J. Econ. Ent. **50**(2):218.

Mackensen, O. 1947. Effect of carbon dioxide on initial oviposition of artificially inseminated and virgin queen bees. J. Econ. Ent. **40**(3):344-349.

. 1948. A new syringe for the artificial insemination of queen bees. Amer. Bee J. **88**(8):412.

. 1954. Some improvements in method and syringe design in artificial insemination of queen bees. J. Econ. Ent. **47**(5): 765-768.

. 1955. Experiments in the techniques of artificial insemination of queen bees. J. Econ. Ent. **48**(4):418-421.

. 1964. Relation of semen volume to success in artificial insemination of queen honey bees. J. Econ. Ent. **57**(4):581-583.

Mackensen, O. and W. C. Roberts. 1948. A manual for the artificial insemination of queen bees. U. S. Bur. Ent. and Plant Quar. Et-250.

. 1952. Breeding Bees. **In:** Insects. U.S.D.A. Yearbook of Agriculture. pp. 122-131.

Mackensen, O. and Kenneth W. Tucker. 1970. Instrumental insemination of queen bees. Agr. Handbook No. 390, U.S.D.A.

Nolan, W. J. 1932a. Multiple matings of the queen bee. **In:** Maryland State Bee Keepers Association Report. 23rd Annual Meeting. pp. 20-34.

. 1932b. Breeding the honey bee under controlled conditions. U.S.D.A. Tech. Bul. 326. 49 pp.

1937a. Improved apparatus for inseminating queen bees by the Watson method. J. Econ. Ent. **39**:700-705.

. 1937b. Bee Breeding. **In:** U.S.D.A. Yearbook of Agriculture pp. 1396-1418.

Oertel, E. 1940. Mating flights of queen bees. Gleanings in Bee Culture. **68**:292-3, 333.

Roberts, W. C. 1944. Multiple mating of queen bees proved by progeny and flight tests. Gleanings in Bee Culture. 72:255-9, 303.

. 1947. A syringe for artificial insemination of honeybees. J. Econ. Ent. 40(3):445-446.

Ruttner, F. 1964. Zur Technik und Anwendung der Künstlichen Besamung der Bienenkönigin. Zeitschrift für Bienenforschung. 7(2): 25-34. April.

. 1968. Insemination artificielle de la reine d'abeilles. Annales de L'Abeille. 11(4):239-319.

, ed. 1969. The Instrumental Insemination of the Queen Bee. Apimondia Publishing House, Bucuresti, I-Str. Pitar Mos., 20, Romania. 78 pp.

, ed. 1975. Die Instrumentelle Besamung der Bienenkönigen. Internationalen Instituts für Bienentechnologie und Wirtschaft der Apimondia. Bukarest, Rumanien. 122 pp.

(ed.). 1976. The Instrumental Insemination of the Queen Bee. International Beekeeping Technology and Economy Institute of Apimondia. Bucharest, Romania. 123 pp.

Ruttner, F., H. Schneider, and J. Fresnaye. 1974. Un appareil standard pour l'insemination artificielle des reines d'abeilles. Apidologie 5(4):357-369.

Taber, Stephen, III. 1954. The frequency of multiple mating of queen honey bees. J. Econ. Ent. 47(6):995-998.

Tucker, Kenneth W., and William C. Roberts. 1968. Honeybees. In: Perry, E. J. (ed.). The Artificial Insemination of Farm Animals. Fourth Revised Edition. Rutgers University Press, New Burnswick.

Tryasko, V. V. 1951. (Signs indicating the mating of queens). Pchelovodstvo 11:25-31. (In Russian). Abstract by M. Simpson in Bee World 34:14, 1953.

Velthuis, H. H. W. and M. J. Sommeijer. 1970. Einige Modifikationen in der instrumentallen Besamung von Bienenköniginnen. Apidiologie 1(3):343-346.

Vesely, V. L. 1965. Die zylinderförmig beendete Inseminationsspitze ermöglicht die künstliche Besamung ohne dass die Sonde gebraucht wird, und die Samenabnahme direkt von den Samenleitern der Drohnen. In: Lucrări Ştiinţifice (1966), 7(1):57-61. (From Ruttner 1975).

Watson, L. R. 1927a. Controlled mating in the honey bee. Rept. State Apiarist, Iowa. pp. 36-41.

. 1927b. Controlled mating of queen bees. Dadant & Sons, Hamilton, Illinois. 50 pp.

Woyke, J. 1960. (Natural and artificial insemination of queen honey bees). Pszczeln. Zesz. Nauk. 4(3-4):183-275 (In Polish). Summarized in Bee World, 1962, 43(1):21-25.

.... 1963a. The behaviour of queens inseminated artificially in different manner. **In:** XIXth International Beekeeping Congress Reports, Prague, pp. 702-703.

.... 1963b. Contribution of successive drones to the insemination of a queen. **In:** XIXth International Beekeeping Congress Reports, Prague, Czechoslovakia, pp. 124-125.

.... 1964. Causes of repeated mating flights by queen honeybees. Jour. Apic. Res. **3**:17-23.

.... 1966. Wovon hängt die Zahl der Spermien in der Samenblase der auf natürlichen wege begatteten Königinnen ab. Zeitschrift für Bienenförschung **8**(7):236-247, April.

.... 1967. Rearing conditions and the number of sperms reaching the queens' spermatheca. XXIst Internal. Apic. Congress. College Park, Md. pp. 232-234.

.... 1971. Correlations between the age at which honeybee brood was grafted, characteristics of the resultant queens, and results of insemination. J. Apic. Res. **10**(1):45-55.

Woyke, J. and Z. Jasinski. 1973. Influence of external conditions on the number of spermatozoa entering the spermatheca of instrumentally inseminated honeybee queens. J. Apic. Res. **12** (3):145-149.

www.ingramcontent.com/pod-product-compliance
Lightning Source LLC
Chambersburg PA
CBHW080616270326
41928CB00016B/3081